International Poultry Library

LEWIS WRIGHT
&
HIS POULTRY

OTHER POULTRY BOOKS

Dr Joseph Batty

Domesticated Ducks & Geese

The Silkie Fowl

Old English Game Bantams

Understanding Modern Game

(with James Bleazard)

Understanding Indian Game

(with Ken Hawkey)

Bantams & Small Poultry

Bantams – A Concise Guide

Poultry Ailments

Sussex & Dorking Fowl

Sebright Bantams

Poultry Characteristics—Tails

Artificial Incubation & Rearing

Natural Incubation & Rearing

Domesticated Ducks & Geese

Indian Runner Ducks

Khaki Campbell Ducks & the Campbells of Uley

The Ancona Fowl

Concise Poultry Colour Guide

Poultry Colour Guide (Large Format)

Japanese Long Tailed Fowl

Poultry Shows & Showing

Natural Poultry Keeping

Practical Poultry Keeping

Understanding Old English Game

Old English Game Colour Guide

The Orpington Fowl

(with Will Burdett)

The Barnevelder Fowl

Marsh Daisy Poultry

Sicilian Poultry Breeds

Rosecomb Bantams

The Orloff Fowl

The Malay Fowl

LEWIS WRIGHT
&
HIS POULTRY

Dr Joseph Batty

Past President: Old English Game Club

Chairman: World Bantam & Poultry Society

NORTHBROOK PUBLISHING Ltd

Beech Publishing House

Station Yard

Elsted Marsh

Midhurst

West Sussex GU29 0JT

ISBN 1-85736-258-6
First published 1983
This Edition 2001

Originally Published as
Lewis Wright's Poultry

British Library Cataloguing-in-Publication Data
A catalogue record for this book is available from the British Library.

NORTHBROOK PUBLISHING Ltd
Beech Publishing House
Station Yard
Elsted Marsh
Midhurst
West Sussex GU29 0JT

Printed by Warwick Printing Company Limited,
Theatre Street, Warwick CV34 4DR.

CONTENTS

	page
Monochorome Illustrations List	vii
List of Coloured Plates	ix

CHAPTERS

1	LEWIS WRIGHT	1
	Early Beginnings	1
	Monograph On the Brahma Fowl	1
	The Practical Poultry Keeper	5
	The Illustrated Book of Poultry	6
	Other Books	8
	Fanciers Gazette	8
	Character and Determination	8
2	THE BREEDS	15
	Anconas	15
	Andalusians	16
	Brahmas	17
	Campines	21
	Cochins	24
	Crêve-Coeurs	29
	Dominiques	29
	Dorkings	30
	Faverolles	33
	Frizzled Fowls	34
	La Flèche	35
	Lakenvelders	35
	Modern Game Fowl	37
	Old English Game Fowl	40
	Hamburghs and Redcaps	46
	Houdans	52
	Modern Langshans	54
	Croad Langshans	56
	Leghorns	58
	Malays, Aseel and Indian Game	63
	Minorcas	69
	Polish (Polands)	72
	Orpingtons	76

	Plymouth Rocks	82
	Rhode Island Red	86
	Scots Dumpies	87
	Scots Greys	87
	Spanish	90
	Sussex	95
	Wyandottes	96
	Yokohamas	104
3	BANTAM BREEDING	107
	Modern Game Bantams	107
	Old English Game Bantams	112
	Variety Bantams	113
4	DUCKS AND ORNAMENTAL WATERFOWL	115
	Ducks	115
	Geese	121
5	GUINEA FOWL AND TURKEYS	127
	Guinea Fowl	127
	Turkeys	128

Light Brahmas

MONOCHROME ILLUSTRATIONS LIST

Figure		*Page*
2.1	Points of the Fowl	13
2.2	Anconas	14
2.3	Silver Campines	23
2.4	Black Cochins	27
2.5	White Dorkings	32
2.6	La Flèche	36
2.7	Lakenvelders	37
2.8	Development of Modern Game	38
2.9	Game Cock Trimmed and Heeled	46
2.10	Hamburghs and Redcaps	47
2.11	Silver Spangled Hamburghs	49
2.12	Derbyshire Redcaps	50
2.13	Black Hamburghs	53
2.14	Langshans	55
2.15	The first or Original Langshans	57
2.16	The Jungle Fowl and some of its Descendants	61
2.17	Black Minorcas	71
2.18	Polands	74
2.19	Black Orpingtons	77
2.20	Rose-comb Rhode Island Red	86
2.21	Scots Greys	88
2.22	Feather from Silkie Hen	89
2.23	Spanish	92
2.24	Sultans	93
2.25	Black Sumatras	94
2.26	Light Sussex Cockerel	96
2.27	Light Sussex Pullet	96
2.28	Wyandottes	100
2.29	Wyandotte Feathers	102
3.1	Sebright Bantams	106
3.2	Various Bantams	108
4.1	Aylesbury and Cayuga Ducks	117
4.2	Toulouse and Embden Geese	122
4.3	African Geese	124
5.1	The Common Guinea Fowl	126
5.2	Wild Turkeys at Home	129

Houdan Head Study

LIST OF COLOURED ILLUSTRATIONS

Between page xvi and 1

Plate 1: Mr H. Leworthy's Pair of Andalusians (1st edition)

Plate 2: Andalusians (**Note: The colour is now the slate grey preferred today**)

Plate 3: American Light Brahmas (1st edition)

Plate 4: Light Brahmas

Plate 5: The Right Hon. Lady Gwydyrs Dark Brahma Cockerel *Sultan* (1st edition)

Plate 6: Mr L. Wright's Dark Brahma Pullet *Psyche* (1st edition)

Plate 7: Dark Brahmas

Plate 8: Mr Julius Sichel's White Cochins *Champion* and *Queen* (1st edition)

Plate 9: Mr E. Tudman's Partridge Cochin *Talbot* (1st edition)

Plate 10: Partridge Cochins

Plate 11: Mr E. Tudman's Partridge Cochin Hen *Titania* (1st edition)

Plate 12: Mr Tomlinson's Buff Cochin Cock *Sampson* (1st edition)

Plate 13: Buff Cochin Hen *Blossom* (1st edition)

Plate 14: Buff Cochins (**Note the deeper buff colour**)

Plate 15: Mr R.B. Wood's Pair of Crêve-Coeurs (1st edition)

Plate 16: Dominiques (1st edition)

Plate 17: Miss Fairhurst's Pair of White Dorkings (1st edition) (**Note the rose comb**)

Plate 18: Mr O.F. Cresswell's Silver-grey Dorking Hen (1st edition)

Plate 19: Silver-grey Dorkings

Plate 20: Mr John Martin's Rose Combed Dorking Cock *Champion* (1st edition)

Plate 21: Mr John Martin's Single Combed Dorking Hen (1st edition)

Plate 22: Dark Dorkings

Plate 23: Faverolles

Plate 24: Mr J.W. Ludlow's Pair of Black Frizzled Fowls (1st edition)

Plate 25: The Hon. W.C.W. Fitzwilliam's Pair of La Flèche (1st edition)

Plate 26: Old English Game: Spangled

Plate 27: Old English Game: Black-red Cock and Clay Hen

Plate 28: Mr John Douglas' Black-breasted Red Game Cock *The Earl* (1st edition)

Plate 29: Mr John Douglas' Black-breasted Red Game Hen *Countess* (1st edition)

Plate 30: Wheaten Game Hen, from a Splendid Bird lent by Mr C. Chaloner (1st edition)

Plate 31: Mr C.W. Brierley's Pair of Brown-Breasted Red Game (1st edition)

Plate 32: Mr John Douglas' Duckwing Game Fowls *Sir Harry* and *Lady* (1st edition)

Plate 33: Mr C.W. Brierley's Pair of Pile Game (1st edition)

Plate 34: Mr John Harris's Pair of Henny Game (1st edition)

Plate 35: Black-breasted Red and Pile Modern Game

Plate 36: Brown Red Modern Game

Plate 37: White and Duckwing Modern Game

Plate 38: The Rev. W. Serjeantson's Pair of Black Hamburgh (1st edition)

Plate 39: Mr Henry Beldon's Pair of Golden-pencilled Hamburghs (1st edition)

Plate 40: Gold Pencilled Hamburghs (**Note the finer pencilling from the earlier illustration**)

Plate 41: Mr H. Pickle's Pair of Silver-pencilled Hamburghs (1st edition)

Plate 42: Mr Henry Beldon's Pair of Golden-spangled Hamburghs (1st edition)

Plate 43: Gold Spangled Hamburghs

Plate 44: Mr Henry Beldon's Pair of Silver Spangled Hamburghs (1st edition)

Plate 45: Mr Robert B. Wood's Pair of Houdans *Young Champion* and *Lady* (1st edition)

Plate 46: Houdans

Plate 47: White Leghorns (1st edition)

Plate 48: Brown Leghorns (1st edition)

Plate 49: White Leghorns

Plate 50: Brown Leghorns

Plate 51: The Rev. A.G. Brooke's Pair of Malays (1st edition)

Plate 52: Malays (front) Aseel (back)

Plate 53: Ayam Jallak, the Finest Breed of Malay Game Cock, drawn from life by a Native Chinese Artist (1st edition). (Note this is not the normal Malay — see text)

Plate 54: Indian Game (Note: type now much changed)

Plate 55: Buff Orpingtons

Plate 56: Plymouth Rocks

Plate 57: Mr P. Unsworth's White-crested Black Polish (1st edition)

Plate 58: Mr Henry Beldon's Pair of Golden-spangled Polish, (1st edition)

Plate 59: Mr Henry Beldon's Pair of Silver-spangled Polish, (1st edition)

Plate 60: Gold Spangled Polish (Polands)

Plate 61: White-faced Black Spanish (1st edition)

Plate 62: The Right Hon. Lady Gwydyr's Pair of Silkies (1st edition)

Plate 63: Mr Robert Loft's Pair of Sultan Fowls (1st edition)

Plate 64: Speckled Sussex

Plate 65: Golden Wyandottes

Plate 66: Silver Laced Wyandottes

Plate 67: Various Bantams (1st edition)

Plate 68: Mr W.F. Entwisle's Black Red & Pile Game Bantams (1st edition)

Plate 69: Mr Matthew Leno's Gold & Silver-laced Sebright Bantams (1st edition)

Plate 70: Modern Game Bantams

Plate 71: Various Bantams

Plate 72: Ornamental Ducks (1st edition)

Plate 73: Mr J.K. Fowler's Pair of Rouen Ducks (1st edition)

Plate 74: (left) Rouen Ducks, (right) Cayuga Ducks

Plate 75: Mrs Mary Seamons' Pair of Aylesbury Ducks (1st edition)

Plate 76: (left) Pekin Ducks, (right) Indian Runner Ducks

Plate 77: (left) Carolina Ducks, (right) Mandarin Ducks

Plate 78: Mr James Watt's Pair of Toulouse Geese (1st edition)

Plate 79: Wild American Turkey Cock (1st edition)

Plate 80: Guinea Fowls, (1st edition) The Property of Mr J.W. Ludlow

Malays (Top) and Aseel

PREFACE AND ACKNOWLEDGEMENTS

Lewis Wright is a name all poultry fanciers know. His books have inspired so many to keep large fowl and bantams for more than a century. I am no exception.

Following an early introduction to poultry keeping as a boy I was so pleased to see and examine *The Illustrated Book of Poultry* by Lewis Wright. At a later stage, when really bitten by the bug, ownership was regarded as vital. In recent years it has been sad to see that many volumes of this beautiful book have been vandalized for the prints which have been sold at quite high prices. Inevitably this has meant that the first edition is becoming scarce and the price is out of the reach of most young fanciers.

In producing the present work the hope is to provide a reference book containing all the main colour plates as well as the historical detail of the breeds.

My thanks are extended to Mrs M E Strange and her son Mr J L Strange (Lewis Wright's grandson) who checked the biographical details and also supplied the wonderful coloured scroll which is produced in monochrome in Chapter 1. All the leaders of the Poultry Fancy sponsored the dinner and presentation which took place in December 1901, providing an appropriate tribute to a dedicated fancier.

This book is presented as a modern reminder of the debt owed to Lewis Wright for his dedication towards an improved fancy.

J Batty

LISS
September 1983.

FOREWORD TO NEW EDITION

Many requests have been made for a new edition of this work which includes the magnificent colour plates painted by the famous artist J W Ludlow.

Many of the standard breeds have changed little and therefore the book gives guidance on what is required for those who are interested in the breeds. How magnificent are these birds, painted to represent the *ideals* which were to be described in the ***Poultry Standards***. Since Lewis Wright had made a special study of these requirements, it is not surprising that they were of such a high standing.

J W Ludlow was also a poultry fancier so he knew what was required. The combined efforts of these two men produced an ever lasting record of the birds developed or imported in those early times, as well as the British breeds, already in existence, which were being improved.

Joseph Batty

Elsted Marsh, December, 2000

LEWIS WRIGHT — journalist, author, fancier, printer, poultry judge and pioneer. (1838-1905).

COLOURED PLATES

The coloured plates in the next section are based on the plates taken from *The Illustrated Book Of Poultry* by Lewis Wright. They are based on paintings by J W Ludlow.

Like most of the work undertaken by the Victorians the paintings show a thoroughness to detail. Moreover, for the first time an attempt was made to show the fancy points of the standard breeds of poultry which could be used as guides for judging.

The first edition plates were produced as chromolithographs and the result was a depth of colour difficult to achieve in later editions. Even the style of painting went through a change.

The second and subsequent editions showed birds which were probably nearer to reality, but the colour and presentation were not so good as in the first edition.

Malays (Top) and Aseel

Plate 1 Mr H. Leworthy's Pair of Andalusians (1st edition)

Plate 2 Andalusians (**Note: The colour is now the slate grey preferred today**)

Plate 3 American Light Brahmas (1st edition)

Plate 4 Light Brahmas

Plate 5 The Right Hon. Lady Gwydyrs Dark Brahma Cockerel *Sultan* (1st edition)

Plate 6 Mr L. Wright's Dark Brahma Pullet *Psyche* (1st edition)

Plate 7 Dark Brahmas

Plate 8 Mr Julius Sichel's White Cochins *Champion* and *Queen* (1st edition)

Plate 9 Mr E. Tudman's Partridge Cochin *Talbot* (1st edition)

Plate 10 Partridge Cochins

Plate 11 Mr E. Tudman's Partridge Cochin Hen *Titania* (1st edition)

Plate 12 Mr Tomlinson's Buff Cochin Cock *Sampson* (1st edition)

Plate 13 Buff Cochin Hen *Blossom* (1st edition)

Plate 14 Buff Cochins (**Note the deeper buff colour**)

Plate 15 Mr R.B. Wood's Pair of Creve-Coeurs (1st edition)

Plate 16 Dominiques (1st edition)

Plate 17 Miss Fairhurst's Pair of White Dorkings (1st edition) **(Note the rose comb)**

Plate 18 Mr O.F. Cresswell's Silver-grey Dorking Hen (1st edition)

Plate 19 Silver-grey Dorkings

Plate 20 Mr John Martin's Rose Combed Dorking Cock *Champion* (1st edition)

Plate 21 Mr John Martin's Single Combed Dorking Hen (1st edition)

Plate 22 Dark Dorkings

Plate 23 Faverolles

Plate 24 Mr J.W. Ludlow's Pair of Black Frizzled Fowls (1st edition)

Plate 25 The Hon. W.C.W. Fitzwilliam's Pair of La Fléche (1st edition)

Plate 26 Old English Game: Spangled

Plate 27 Old English Game: Black-red Cock and Clay Hen

Plate 28 Mr John Douglas' Black-breasted Red Game Cock *The Earl* (1st edition)

Plate 29 Mr John Douglas' Black-breasted Red Game Hen *Countess* (1st edition)

Plate 30 Wheaten Game Hen, from a Splendid Bird lent by Mr C. Chaloner (1st edition)

Plate 31 Mr C.W. Brierley's Pair of Brown-Breasted Red Game (1st edition)

Plate 32 Mr John Douglas' Duckwing Game Fowls *Sir Harry* and *Lady* (1st edition)

Plate 33 Mr C.W. Brierley's Pair of Pile Game (1st edition)

Plate 34 Mr John Harris's Pair of Henny Game (1st edition)

Plate 35 Black-breasted Red and Pile Modern Game

Plate 36 Brown Red Modern Game

Plate 37 White and Duckwing Modern Game

Plate 38 The Rev. W. Serjeantson's Pair of Black Hamburgh (1st edition)

Plate 39 Mr Henry Beldon's Pair of Golden-pencilled Hamburghs (1st edition)

Plate 40 Gold Pencilled Hamburghs (**Note the finer pencilling from the earlier illustration**)

Plate 41 Mr H. Pickle's Pair of Silver-pencilled Hamburghs (1st edition)

Plate 42 Mr Henry Beldon's Pair of Golden-spangled Hamburghs (1st edition)

Plate 43 Gold Spangled Hamburghs

Plate 44 Mr Henry Beldon's Pair of Silver Spangled Hamburghs (1st edition)

Plate 45 Mr Robert B. Wood's Pair of Houdans *Young Champion* and *Lady* (1st edition)

Plate 46 Houdans

Plate 47 White Leghorns (1st edition)

Plate 48 Brown Leghorns (1st edition)

Plate 49 White Leghorns

Plate 50 Brown Leghorns

Plate 51 The Rev. A.G. Brooke's Pair of Malays (1st edition)

Plate 52 Malays (front) Aseel (back)

Plate 53 Ayam Jallak, the Finest Breed of Malay Game Cock, drawn from life by a Native Chinese Artist (1st edition). (Note this is not the normal Malay — see text)

Plate 54 Indian Game (Note: type now much changed)

Plate 55 Buff Orpingtons

Plate 56 Plymouth Rocks

Plate 57 Mr P. Unsworth's White-crested Black Polish (1st edition)

Plate 58 Mr Henry Beldon's Pair of Golden-spangled Polish, (1st edition)

Plate 59 Mr Henry Beldon's Pair of Silver-spangled Polish, (1st edition)

Plate 60 Gold Spangled Polish (Polands)

Plate 61 White-faced Black Spanish (1st edition)

Plate 62 The Right Hon. Lady Gwydyr's Pair of Silkies (1st edition)

Plate 63 Mr Robert Loft's Pair of Sultan Fowls (1st edition)

Plate 64 Speckled Sussex

Plate 65 Golden Wyandottes

Plate 66 Silver Laced Wyandottes

Plate 67 Various Bantams (1st edition)

Plate 68 Mr W.F. Entwisle's Black Red & Pile Game Bantams (1st edition)

Plate 69 Mr Matthew Leno's Gold & Silver-laced Sebright Bantams (1st edition)

Pile

Brown Reds

Silver
Duckwings

Black Reds

Plate 70 Modern Game Bantams

Plate 71 Various Bantams

Plate 72 Ornamental Ducks (1st edition)

Plate 73 Mr J.K. Fowler's Pair of Rouen Ducks (1st edition)

Plate 74 (left) Rouen Ducks, (right) Cayuga Ducks

Plate 75 Mrs Mary Seamons' Pair of Aylesbury Ducks (1st edition)

Plate 76 (left) Pekin Ducks, (right) Indian Runner Ducks

Plate 77 (left) Carolina Ducks, (right) Mandarin Ducks

Plate 78 Mr James Watt's Pair of Toulouse Geese (1st edition)

Plate 79 Wild American Turkey Cock (1st edition)

Plate 80 Guinea Fowls, (1st edition) The Property of Mr J.W. Ludlow

CHAPTER 1

LEWIS WRIGHT

When Lewis Wright was accidently killed whilst crossing the railway line to catch a train, he had been well known and acclaimed for his knowledge of poultry for more than 30 years. His monumental work *The Illustrated Book of Poultry* had been in print for that period and he had been active in poultry matters before then. Indeed, he had taken an interest from the early age of nine.

EARLY BEGINNINGS

Born in 1838 Lewis Wright, the son of a well-known printer and publisher, appeared destined for a career in the world of books. This, in fact, is what happened because he went into printing and learned the trade. Indeed, until marrying at the age of twenty-six, his initial foray into poultry-keeping described below had been allowed to lapse.

As noted, his interest in poultry started at the early age of nine. In 1847, with his brother, he was allowed a breeding pen of Minorcas consisting of a cock and three hens. These were kept in a small back yard and turned out to be splendid layers. Those early days appeared to lay a foundation of what was to become a dedicated occupation in later years.

At school, usual activities continued and he and other boys were allowed to keep birds in a hen roost constructed in the cellar of the headmaster's house. This joint effort in co-partnership was not a great success and Lewis found himself rather dissatisfied with the venture. He did, however, continue with his general interest in poultry and attended poultry shows which were now being established with great enthusiasm by fanciers. The spirit of wanting to belong to the poultry Fancy and to recognise the merits of well-bred birds was felt even in these early days. S.H. Lewer, a close friend and colleague for many years, wrote in *The Feathered World* that Mr. Wright had told him of the admiration he felt when he attended gatherings of poultry fanciers in those far off days when the first interest in exhibition poultry was being shown.

There was a period when Lewis Wright was too busy learning about printing to have time for poultry keeping. In addition, he studied scientific subjects, an area which was to interest him in later life to the extent of writing a book on the study of light. The fact remains that, after a period when he did not keep poultry, he became a dedicated fancier. According to S.H. Lewer, the interest arose when he married at the age of twenty-six and found that his wife was interested in keeping a few birds for eggs. From this new beginning, Lewis Wright became a dedicated fancier keeping many breeds, but his main interest was in Brahmas. As a result his work *The Brahma Fowl* was published, appearing in 1870 and followed eighteen months after by a second edition.

Lewis Wright combined many qualities and activities into an extremely full business life. He was a journalist, author, fancier, printer, editor, poultry judge and, above all, a pioneer who sought to develop and improve the poultry Fancy. The poultry shows were being taken seriously but there was need for rules and standards without which there could be no general acceptance of the awards given by judges. Some of his achievements are considered below.

THE MONOGRAPH ON THE BRAHMA FOWL

In the second edition of his work on the Brahma Fowl, Lewis Wright expressed surprise that this new edition had been called for within eighteen months of the first edition. Moreover, the number of Brahmas being shown at the principal shows had doubled. Clearly the existence of expert guidance had encouraged new fanciers and, although other poultry books were being published and great controversy centred around the breed, there is no doubt that Lewis Wright's work was well received. The enthusiasm and persuasion which signpost the book were bound to result in the creation of more followers. He regarded the Brahma in poultry as the same as the Shorthorn in cattle. It would thrive and be an economic proposition under most, though not all, circumstances. As a bird for the table it had excellent properties and it was found to be an excellent layer, around 150 eggs per annum being feasible. Even today for a fancy (standard bred) fowl, this average would be regarded as commendable.

When chicks are required, the hen proves to be a diligent mother. In fact, Lewis Wright advocated that she should be allowed at least one brood per annum. This allowed her to follow her natural instincts and reduced the danger of accumulating internal fat, which is a problem of all feather-legged breeds.

The monograph extended to just over 140 pages and included four plates in colour. The specimens illustrated were not as "feathery" as the modern species or even as the birds illustrated in *The Illustrated Book of Poultry*. The book presented a detailed analysis of the origins of the Brahma Fowl. He came to the conclusion that the Dark and Light varieties came from distinct origins. Furthermore, he criticised at great length the claims of G.P. Burnham who had written a book entitled *The History of the Hen Fever* (1855).

Another great authority of the period, Harrison Weir, who produced the massive work *Our Poultry* did not always agree with Lewis Wright and so on this issue of The Brahmas he appears to side with Burnham. He refers to the original name, the American Chittagong or Shanghai, later to be called the Brahma Pootra. In the year 1852, Mr. Burnham sent nine birds to Her Majesty Queen Victoria and they were called Grey Shanghais. Harrison Weir, a fine artist, was commissioned to draw them for the *Illustrated London News* and he also painted seven of the fowls in a group.

Whoever was correct on the precise origin of the Brahma varieties will probably never be known. The fact remains that Lewis Wright was very successful with his Dark Brahmas winning at the Crystal Palace and at Birmingham. Moreover, from his experience of breeding, exhibiting and judging, as well as obtaining the opinions of other judges, he was able to formulate methods of judging by a points method which subsequently led to a universal system being developed. His words are reproduced below:

On the Judging of Brahmas

There is no question whatever that the publication of the well-known *Standard of Excellence in Exhibition Poultry*, drawn up as it was by the best breeders and judges of the day, did much to promote a greater uniformity of judging both in Brahmas and other fowls; and whatever the fault be found with it, we think, on the whole, it will ever remain the basis at least of any future system of judging. It in fact opened up an entirely new question, as to whether the judging of poultry *could* be conducted upon any "system" at all — a matter not altogther settled even now. It was contended by many that correct judging could only be performed by "the eye" of an experienced individual, whose general opinion of a bird would be of greater value than any conclusion arrived at by other means. On the other hand, the promoters and compilers of *The Standard* proposed to substitute for this — which may be called the empirical method — a regular system of deciding by "points", each of which was to have a certain numerical value, and which were to be added up to determine the relative value of a bird, much in the same way as the well-known "good marks" in schools. To say that no judge could go about "books in hand" and actually "add up the points" of the best birds, is not to offer any valid objection to such a system, since long experience and habit would give to him a ready and instinctive appreciation of the value of the points, which would amount to nearly the same thing; and the experiment was accordingly regarded with great interest by all intelligent poultry breeders.

With regard to the actual result opinions differ: but for ourselves, as we have already said, we think there has been of late years an appreciable and gratifying improvement in the uniformity of judging, clearly traceable to *The Standard;* whilst on the other hand, we have repeatedly had occasion to observe how in certain cases the system completely breaks down. The question then naturally arises, whether in such cases of failure the fault is inherent in the system itself, or arises from insufficient or erroneous development of it. This question is a very interesting one, but we cannot discuss it here, being foreign to the special subject of these pages. We will only say, that a new system could not be expected to reach perfection at the first attempt; and that so far as Brahmas are concerned, while we feel convinced, after many attempts to frame a perfect schedule of points for them, that no theory can *entirely* dispense with what we have called the instinct or general impression of the judge, we do think a very nearly perfect system may be devised, and such as will in case of doubt solve almost any practical difficulty. In the few remarks which follow, therefore, we do not presume to teach authorities already eminent how to judge a Brahma, but simply to assist those who may be conscious they do not thoroughly understand this particular breed, and who may have found by experience that to literally follow *The Standard* in all cases occasionally leads to manifest errors. We can illustrate our meaning best by first giving the value of points in Brahmas as laid down in the work referred to, where they stand thus:

POINTS OF BRAHMAS

Size	3
Colour	4
Head and Comb	1
Wings. Primaries well tucked under secondaries	1
Legs, and feathering of ditto	1
Fluff	1
Symmetry	2
Condition	2
Total	15

DISQUALIFICATIONS.— Birds not matching in the pen; combs not uniform in the pen, or falling over to one side; crooked backs; legs not feathered to the toes, or of any colour except yellow or dusky yellow.

The system implied here is easily understood. A pullet perfect in colour will count 4 on that ground; but if not perfect will count only 3, 2, or 1, according to the judge's opinion of her marking; or if he thinks her very bad will lose 4 points. If wanting in leg-feather, however, she will only lose 1. Thus all her points are added up, and by the result she is to be compared with other birds. If, however, a bird has a very bad falling comb, it is to be disqualified; and the same of a crooked back and some other faults.

Now applying this to the decisions of the best judges, it will be found that while often pretty consistent, many evidently *correct* decisions do not harmonize with it. For instance, we well remember Mr. Hewitt once awarding a cup to a cockerel with a very large shapeless comb hanging considerably over, and legs almost white; on both of which accounts, according to the above, the bird should have been disqualified. We are not sure we should ourselves have awarded him the cup; but we should without hesitation have given him the first prize (a position equally inconsistent with *The Standard)* the reason being, that the bird was so rarely perfect in every other point, no judge could possibly have passed him over. Again, let us suppose a cockerel to lose one point by not being very large, another by a little want of condition, and another by some degree of fault in the comb: if perfect in other matters his numerical value would be 12. The bird in the next pen might carry every point with the sole exception of symmetry, in which he was altogether wanting, being clumsy in shape and most ungainly in carriage: this fault would lose him 2 points, and his value would be 13, beating the other by one point. But we have not the least hesitation in saying that nearly any judge — certainly any *good* judge — would exactly reverse such a conclusion. Many other similar cases might be given, and are constantly occurring.

It will be seen, then, that the primary fault of *The Standard* system is that it is not *elastic* enough: and after many trials to amend it, we feel persuaded that 15 points are *not sufficient* to form any reliable "standard" at all of some breeds; whilst there are several elements unmentioned, which ought to be included in the calculation. Respecting some points, also, opinion has perceptibly changed since the original tables were published. We have therefore ventured to suggest an extended scale, not as our own opinion merely, but as embodying what careful comparison has taught us is about the value now *practically* given to various points, as evidenced by the best actual *decisions* of Messrs. Hewitt and Teebay during the last four years. The difficulty of tabulating such actual practice and reducing it to a definite system, none but those who have attempted it can estimate; and we offer the subjoined scale with some diffidence, as another attempt to enable judges of less experience to arbitrate in harmony with the practice of the best recognized authorities.

It will be seen that in our table of points we have separated the sexes. We have done so because all our observations have led us to the conclusion that to apply the same scale to both is to necessitate a false standard for one or the other.

In our opinion, then, the general table of values should stand as follows. The italics represent our personal views only: all the rest is reduced from actual analysis of the value now practically given to the points, any decisions apparently erroneous having been excluded from the computation.

THE COCK

POINTS OF MERIT		SPECIAL DEFECTS	
		To count against the bird in proportion to their degree.	
Size	4		
Colour	4	Stain of white in deaf ear	1
Smallness, shape, and expression of Head	1	White legs	3
Comb	2	Primaries of wing not tucked in	3
Fullness of Hackle	1	*Vulture hocks*	3-4

Wings, proper size and position of	1
Legs and feathering	2
Fluff	1
Breadth of Saddle	1
Rise of ditto	1
Tail	2
Symmetry	2
Condition and handsome appearance	3
	25

White in the tail	3

DISQUALIFICATIONS

Round or crooked back, crooked beak, or any bodily deformity; knock-knees, or any fraudulent dressing or trimming.

With reference to the table of defects, it should be observed that only those can manifestly be inserted which are not provided for in the points of merit. Thus, smallness would be a great fault; but the bird would according to the points of merit lose four by want of size, and no other provision is therefore necessary. So of defects in comb and feathering. A tail too long or sweeping on the one hand, or too much like a Cochin on the other, would also lose a bird either one or two points. Hence the table of defects is only needed for special faults which in practice are found not sufficiently accounted for in the general scale.

In regard to the special points of difference, it will be seen that, amongst other things, we have given much greater weight than *The Standard* to disorder of the wings. Like the other points wherein we differ, this results from comparison of many decisions by the authorities named; who — we think quite rightly — have on all late occasions laid great stress on the fault in question, so much so that in good classes a bird is now almost disqualified unless he carries his wings clean. Comb and leg-feathering are also now given, in practice, at least the values we have attached to them.

In suggesting that nearly three points against a bird should be allotted to white in the tail, we differ somewhat from one or two judges for whose opinion we have the highest respect; though we believe every judge of repute attaches more or less weight to the fault in question. In urging that so much stress should be laid upon this fault, we certainly are influenced partly by the fact that Brahmas are in a great degree birds of feather; and that if so, such a glaring blemish ought to be regarded as in the case of a Silver-grey Dorking. But a far stronger reason is that hardly any fault has so strongly hereditary a character, which makes it most desirable to stamp out. We also believe it to be often a sign of former Dorking taint; for we have very rarely found it except in strains which had large coarse heads, a point we have already mentioned as indicating Dorking blood. All these reasons, therefore, demand that more value should be attached to purity of colour in the tail than has hitherto been the case, if it be desired to secure even purity of breed.

With regard to colour, we would simply express our opinion that a bird of the objectionable straw colour we have already referred to, should lose the full number of four points; the fault being lately so much increased as to need this decided check.

We now pass to the hen

THE HEN

Points of Merit

Size	3
Beauty and regularity of colour and marking	4
Smallness and beauty of head	2
Comb	1
Shortness and breadth of back	1
Cushion	2
Fluff	1
Legs and feathering	2
Shape	2
Condition, carriage, and general appearance	2
	20

SPECIAL DEFECTS

White legs	2
Very long tail	2
Primaries not tucked in	3
Very streaky feathers, though otherwise good color (in Dark)	2
Shank feathering not pencilled as body (in Dark)	1
Spotted back (in Light)	2
Vulture hocks	2-3

DISQUALIFICATIONS

Round or crooked backs; crooked bills; knock-knees, or any bodily deformity; large red or white splashes in the Dark breed; pinky legs, or any fraudulent dyeing, dressing, or trimming.

In the hen, it will be observed, we have attached more value to the head and less to the comb than in the cock. This is in conformity both with all recent decisions, and with strict propriety; as are the other variations from or additions to the numerical values given for the male bird.

The above scales we think, after much comparison and testing, will be found to harmonize with the practice of the two eminent judges we have named, in almost every case correctly judged. We feel convinced that to form any system at all, it is needful to value marked *defects* as well as points of merit; and that it is from the want of this that *The Standard* has not been found comprehensive enough when tested by actual results. It must in fairness be admitted, however, that the Brahma — especially the Dark Brahma — is one of the most difficult breeds of all to do with; and that no system can possibly dispense with what we may call the *instinct* of a good judge. Mr. Hewitt will often, as we have witnessed, pick out the first-prize bird by one glance of his eye down the class, more correctly than a less experienced arbitrator would do after an hour's examination with the most perfect "scale of points" in his hand. No such system can in fact *make* a judge, where this natural "eye" for a fowl is altogether wanting; but it may still be of great use, especially in judging breeds with whose special points the arbitrator is less acquainted; and may also serve, in the rare cases where a good judge has — from want of time, nervousness, or sheer fatigue of mind — made a wrong decision, to prevent exhibitors being misled by such an error into following a really faulty standard. It is with such objects only that these remarks are offered for the consideration of those who have occasion or desire to study the correct judging of the Brahma fowl.

It is absolutely necessary to notice in this place the vexed question regarding vulture-hocks. *The Standard of Excellence* pronounced these to be "objectionable, but *not* a disqualification". Since then the hock has been absolutely disqualified, with two evident consequences. In the first place, breeders acquired such an absurd *dread* of disqualification for hock that they feared to breed even well-feathered birds: some judges would not look at even a *moderately* feathered shank, and nearly "bare poles" became the ordinary fashion. The ugliness of this, however, speedily caused most exhibitors to give it up in sheer disgust; with the second consequence that many who would not *wait* to breed the lost feather back fairly, bred hocked birds, plucked the hock, and showed them thus to win prizes. We will not mention names — these pages being written to assist, and not to wound others — but truth compels us to state from personal observation, that this was done and prizes were so won in *dozens of cases;* and we have several times heard the trickery confessed and justified by the offender's own lips. On every such occasion the ground of justification has been, that the hocked specimens have been amongst the "best birds", and it has often been added, "we can't afford to give them up, sir".

Now as the only object of disqualifying any particular feature is to banish it from the fancy, and since the disqualification of vulture-hock has altogether failed in this effect; since hocked birds *are* shown — fraudulently — and being then purchased disgust many purchasers who do not like the appendage; since many experienced breeders affirm that it is often the best specimens in other respects which are so furnished, and refuse to give them up; and since it is found in practice that the time allotted for judging is not enough to detect fraud in very many cases; from all these considerations it appears that absolute disqualification has not only missed its sole possible end, but has actually produced serious evils. We cannot resist the conclusion that under these circumstances the dictum of *The Standard* should be again regarded as the rule of judging, and that vulture-hocks should be considered "objectionable, but *not* a disqualification", that is, that the hock should weigh sensibly against a bird; but that a specimen of marked superiority in every other point should not be disqualified, but be judged as *whole.* We object to the hock ourselves as most unsightly, but this conclusion is forced upon us; we believe it is becoming the conviction of many judges themselves; and we appeal to them, as having studied this particular fowl with ever-increasing interest for several years, to hasten by their decisions and all other means in their power the general adoption of such a rule.

In our opinion vulture-hocks should have the value of 3 or 4 points against a cock, and 2 or 3 points against a hen. Less than this would not attain the desirable end of discouraging the hock as much as possible; while more would still hold out temptation to fraudulent proceedings, and amount in practice to the disqualification which has wrought such evil effects.

We have frequently seen Brahma cockerels much plucked about the tails. This is usually done to produce the effect of the Cochin tail, which so many seem ignorantly to aim after; and the effect is always bad — far worse than nature — to any one who understands fowl. We need not say that this or any other fraudulent proceeding should subject a pen to instant and ignominious disqualification.

Finally, we would repeat that there has been of late an increasing tendency to show birds *too fat.* All large breeds are subject to this evil; and years ago, when it was even more prevalent than now, Mr. Hewitt did good service by passing over in a marked manner many over-fattened hens, and thereby discouraging the practice of feeding show fowls to the highest point, to their utter ruin. Again, however, the same vicious system appears to be gradually creeping in; and as the practice is really dishonest, it should be checked as far as possible by never awarding prizes to hens which evidently carry more fat than is consistent with real health and condition.

THE PRACTICAL POULTRY KEEPER

This general work on poultry keeping was a best seller by any standards. Around 1866, Lewis Wright recognised the need for a practical and authoritative book on poultry keeping. That he was right is evidenced by twenty editions and sales approaching eighty thousand copies — surely a great achievement for a specialised work.

Despite its popular appeal, this title never achieved the accolade given to *The Illustrated Book of Poultry,* the mammoth work which appeared later and which was an expansion of his earlier book. *The Practical Poultry Keeper* examines all the essential requirements for keeping birds healthy. The twenty-eight chapters in the twentieth edition extend from the essentials of management, followed by breeding and exhibiting, then the different breeds, and, finally, turkeys, ornamentals and waterfowl. The eight coloured plates by J.W. Ludlow enable the readers to appreciate the beauty of standard bred poultry.

The book is essentially practical. Feeding soft food early in the morning is advocated, thus enabling a fresh supply to get into the system and not merely into the crop (hard corn would require to be digested in the gizzard). There are many other hints on the *how* and *why* of poultry keeping. Undoubtedly, it was this approach which made the book so popular, aided by the concise, readable style of the accomplished writer. Many of the later books copied the approach adopted by Lewis Wright and, therefore, he was a major influence on the writers who followed him on poultry subjects.

THE ILLUSTRATED BOOK OF POULTRY

This book arrived at a most opportune time, when interest in poultry-keeping as a hobby was at its height. Developments, changes, improvements, the introduction of new breeds, royal patronage and the growing interest in the many and fascinating aspects of a splendid hobby all helped to make the work of Lewis Wright and his co-contributors extremely welcome.

The contents of *The Illustrated Book of Poultry* relating to the different breeds are reproduced later in the book, where appropriate with additional notes. Here it is important to note the two important features which call for special mention:

1. Portraits by J.W. Ludlow

According to Lewis Wright, Mr J.W. Ludlow co-operated with him to the full, 'striving to meet our wishes in every point, and in many cases even deliberately sacrificing more pictorial "effect" in order to bring out the *points of the birds,* which he knew to be the object we chiefly desired.'

In everyday language, Ludlow painted to portray the ideal type of bird and not necessarily birds which actually existed. These served to guide fanciers in breeding top quality birds.

2. Schedules for judging with Scales for Guidance

These were the forerunners of the first poultry *standards* and from this beginning there developed the official *Poultry Club Standards.* An example of a schedule is reproduced below:

SCHEDULE FOR JUDGING BRAHMAS

GENERAL CHARACTERISTICS OF COCK – *Head and Neck* – General appearance of head very short, small and intelligent; beak short, curved, and stout at the base; comb triple or in three ridges, resembling three small combs, the centre being the highest, and the whole small, low, and firm on the head, the centre ridge *perfectly* straight and neatly serrated; wattles moderately long, thin, and pendant; deaf-ears large, and hanging below the wattles; neck well-proportioned and finely curved, as in a spirited horse, and very thickly furnished with long hackle, which should flow well over back and shoulders. *Body* — General shape large and deep, but tight and compact in make; back broad and short; saddle very broad and large, with a gradual and decided rise to the tail, so as to form no angle with that member; wings larger than in Cochins, but still small and neatly tucked up, with secondaries carried well under the primaries; breast full, prominent, and reaching well down. *Legs and Feet* — Thighs large and well-furnished with fluffy feathers, the hocks being entirely covered with soft curling feathers, but free from stiff quills (vulture-hock) which are particularly objectionable; shanks rather but not too short, thick, wide apart, and heavily feathered down the outside, the feathering to start out well from the hock, and continue to ends of outer and middle toes; toes large, straight, and well spread out. *Tail* — Much larger than in Cochins, but still small, carried nearly but not quite upright, and the top pair of feathers curving outwards as in the tail of the black-cock; sickles very short and not curving downwards, but lesser sickles and tail-coverts very abundant, covering nearly the whole sides of the tail. *Size* — Very large, ranging from eleven pounds to fifteen pounds in cocks, and eight pounds to eleven pounds in cockerels. *General Appearance* — Very symmetrical and compact. *Carriage* — Noble and commanding, with the head carried very high.

GENERAL CHARACTERISTICS OF HEN — *Head and Neck* — General appearance of head, very small, peculiarly arch, and intelligent, caused by a slight fulness over the eye, which should on no account tend to coarseness; beak and head rather short, as in the cock; comb as small as possible, a large loose comb being particularly objectionable; deaf-ears well developed; wattles nicely rounded, neat, and free from any folds; neck short, very full in hackle, and free from twist in the hackle. *Body* — General shape square and neat; back wide, flat across, and short, cushion broad and large, not convex or globular as in Cochins, but rising to the tail; wings moderate in size, and well tucked into the fluff and cushion-feathering; breast very prominent, low down, and full. *Legs and Feet* — As in the cock, but as short as possible. *Tail* — Rather short, so as not to rise much above the extremity of the cushion, and carried very nearly upright. *Size* — Very large, ranging from eight pounds to thirteen pounds in hens, and six pounds to nine pounds in pullets. *General Appearance* — Massive and square, but neat and compact. *Carriage* — Matronly and dignified, both head and tail being well carried up.

COLOUR OF LIGHT BRAHMAS — *In both Sexes* — Beak a rich yellow, with or without a dark stripe. Comb, face, deaf-ears, and wattles brilliant red, with as few spiky feathers as possible. Eyes pearl or red, red being preferable. Shanks a brilliant orange-yellow. *Colour of Cock* — Head silvery white; hackle white, striped with black as distinctly as possible; saddle-feathers either white or white lightly striped with black; tail and tail-coverts glossy green-black, except the two top feathers, which may or may not be laced with white. Rest of the body a pearly surface-colour, with grey under-fluff seen when the plumage is ruffled; the secondaries being white on lower edges and black on the inner, and primaries black. The shank-feathering white more or less mottled with black. *Colour of Hen* — Head silvery white; hackle white, heavily striped with bright intense black; tail black, except the top pair, which should be edged with white. Rest of the plumage white on the surface and grey in the under fluff, with wings and leg-feathers as in the cock. (American fanciers admit a creamy tinge to the white; but perfect birds should be of a pure and pearly colour all over).

COLOUR OF DARK BRAHMAS — *In both Sexes* — Beak yellow, yellow with a dark stripe, horn-colour, or black. Eyes pearl or red, the latter preferable. Comb, face, deaf-ears, and wattles brilliant red, as little obscured by feathers as possible, the beard, or feathers under the throat, not exceeding moderation. *Colour of Cock* – Head silvery white; hackle white, heavily and sharply striped with rich black, as free from white streak in the centre as possible. Saddle-feathers the same. Back and shoulders silvery white, except between the shoulders, where the feathers should be black laced with white. Upper wing-butts black; bow silvery white; bar, or coverts, glossy black "shot" with green; secondaries white on outside web, which is all that appears when wing is closed; black on inside; the end of every feather black. Primaries black except a narrow white edge on outside web. Breast, under parts, and leg-feather glossy black, as intense as possible, or black evenly and sharply mottled with small spots of white. Fluff black, or black laced or tipped with white. (All black in the under parts preferable for exhibition). Tail black, richly "shot" or glossed with colour — white not a disqualification, but very objectionable. Shanks a deep orange-yellow. *Colour of Hen* — Head and hackle silvery white, heavily and sharply striped with black, the marking to extend well over the head. Tail black, the top pair edged with grey. Rest of the plumage a silver-grey, dull grey, or steel-grey ground-colour, accurately pencilled over in a crescentic form with steel grey, blackish grey, or black; the breast to be perfectly marked, and free from streaks up to the throat; a chestnut tinge not objectionable if of a rich and not dingy character. The leg-feather to be pencilled as the body. Shanks deep yellow, with or without a dusky tinge.

VALUE OF DEFECTS IN JUDGING

Standard of Perfection

A bird ideally perfect in shape, size, colour, head and comb, cushion or saddle, leg-feathers, tail, &c., and in perfect health and condition to count in points.	100
If of extraordinary size, add on that account	5

Defects to be Deducted

Bad head and comb (comb to count 7 in cocks and 5 in hens)	12
Scanty hackle	5
Want of cushion	7
Want of fluff	6
Want of leg-feather	7
Vulture-hocks	20
Bad shape or carriage of tail	6
White in tail	10
Primaries out of order*	15
Pale legs	8
Curved toes	7
Stain of white in deaf-ear	5
Splashed or streaky breasts in Dark, or black specks in Light	12
Shank-feather (in Dark hens) not pencilled as the body	4

Other faults of colour	10
Want of size	20
Want of general symmetry	15
Want of condition	12
Want of condition (if total)	35

**This refers to primaries merely "slipped" outside the wing. For primaries actually twisted on their axes, see list of disqualifications below.*

DISQUALIFICATIONS — Birds not tolerably matched. Primary feathers twisted on their axes. Utter absence of leg-feather. Pinky legs. Large red or white splashes in Dark birds, or conspicuous black spots in Light. Round or crooked backs, wry-tails, crooked bills, knock-knees, or any bodily deformity. Any fraudulent dyeing, dressing, or trimming.

In the eighth edition of *The Poultry Club Standards of Perfection* edited by Wm. Broomhead in 1930, the debt to Lewis Wright is acknowledged. The first *Standards* appeared in 1865, but these lapsed and were reconstituted in 1886. Undoubtedly the schedules appearing in *The Illustrated Book of Poultry* (first edition 1874) were used as a basis. However, his earlier system of having points for defects was modified into points for specified requirements such as **Type, Colour, Head, Legs** and so on.

Lewis Wright played the major part in getting the *standards* accepted. He had observed the judging at poultry shows and had seen the malpractices which could occur. He discussed the methods and the best ways of introducing a scientific approach to awarding prizes. Sometimes in his endeavours he made himself unpopular. Feather plucking and other malpractices were quite common features and Lewis Wright set out to eliminate them. The foundations he established live on today and any faking is dealt with quite severely.

OTHER BOOKS

Although poultry received much attention, other hobby activities were not neglected. The book *The Practical Pigeon Keeper,* beautifully illustrated with drawings by Ludlow, covered all the essentials for managing, breeding, exhibiting and enjoying Fancy pigeons as a hobby. Before then he had edited *The Book of Pigeons* by R. Fulton, the most important book on pigeons ever to be published and illustrated with fifty paintings by J.W. Ludlow in the same format as *The Illustrated Book of Poultry.* He also turned his attention to the editing of Blakston's *Canaries and Cage Birds.* All this work was for Cassell, Petter, Galpin & Co. (later Cassell & Co.) for whom he worked as an editor until 1900 when he returned to Bristol to join the firm of Messrs. John Wright & Co. as the chief editor.

THE FANCIERS' GAZETTE

Lewis Wright was the editor of *The Fanciers' Gazette,* a magazine founded in 1874. This was expanded and renamed *The Livestock Journal and Fanciers' Gazette* and he continued as editor until 1880. Under his guidance and control, journalism in poultry keeping reached a high level never before achieved. He used the journal as a platform upon which to voice his views and to try to improve the poultry Fancy.

CHARACTER AND DETERMINATION

Lewis Wright was regarded by his contempories as a man of high ideals who was prepared to fight for a cause. At the same time, being a rather reserved individual, he did not deliberately seek publicity for its own sake. In fact, in his later years although his interest in poultry continued and he was still an active writer, he no longer exhibited or judged at shows. *The Illustrated Book of Poultry* was reissued, later revised and was then modified and published in another form. This and similar work including the *Poultry Standards* kept him very busy.

In honour of his work a special dinner was arranged where all the leading personalities assembled. The illuminated scroll which was presented to Lewis Wright is probably unique and for this reason the words are reproduced below:

To Lewis Wright, Esq.

In asking you to accept this Address, together with the enlarged photograph of yourself, we, voicing the great community of breeders of poultry, desire, on the occasion of your departure from London and the approaching completion of the Revised Edition of your great work, "The Book of Poultry," to express our sense of the indebtedness felt by poultry-breeders of every class, both by the amateur or professional exhibitor, and by those who devote their attention to the wide interests concerned in food supply, with regard to your labours, and especially in respect to your writings, which are read and studied throughout the Civilised World.

We rejoice in the fact that, though you are personally unknown to many of the younger generation of poultry-breeders, the last Edition

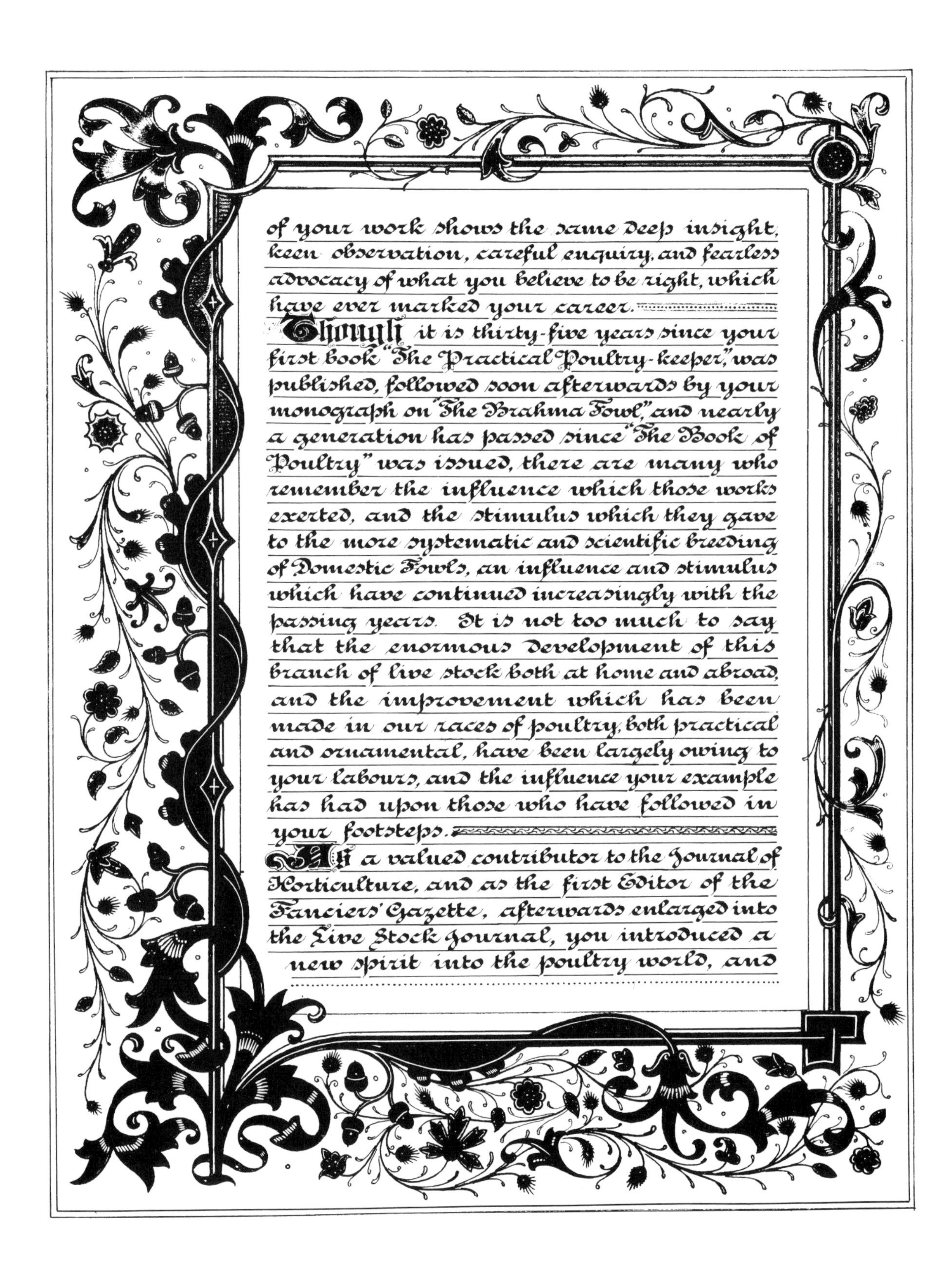
of your work shows the same deep insight, keen observation, careful enquiry, and fearless advocacy of what you believe to be right, which have ever marked your career.

Though it is thirty-five years since your first book "The Practical Poultry-keeper," was published, followed soon afterwards by your monograph on "The Brahma Fowl," and nearly a generation has passed since "The Book of Poultry" was issued, there are many who remember the influence which those works exerted, and the stimulus which they gave to the more systematic and scientific breeding of Domestic Fowls, an influence and stimulus which have continued increasingly with the passing years. It is not too much to say that the enormous development of this branch of live stock both at home and abroad, and the improvement which has been made in our races of poultry, both practical and ornamental, have been largely owing to your labours, and the influence your example has had upon those who have followed in your footsteps.

As a valued contributor to the Journal of Horticulture, and as the first Editor of the Fanciers' Gazette, afterwards enlarged into the Live Stock Journal, you introduced a new spirit into the poultry world, and

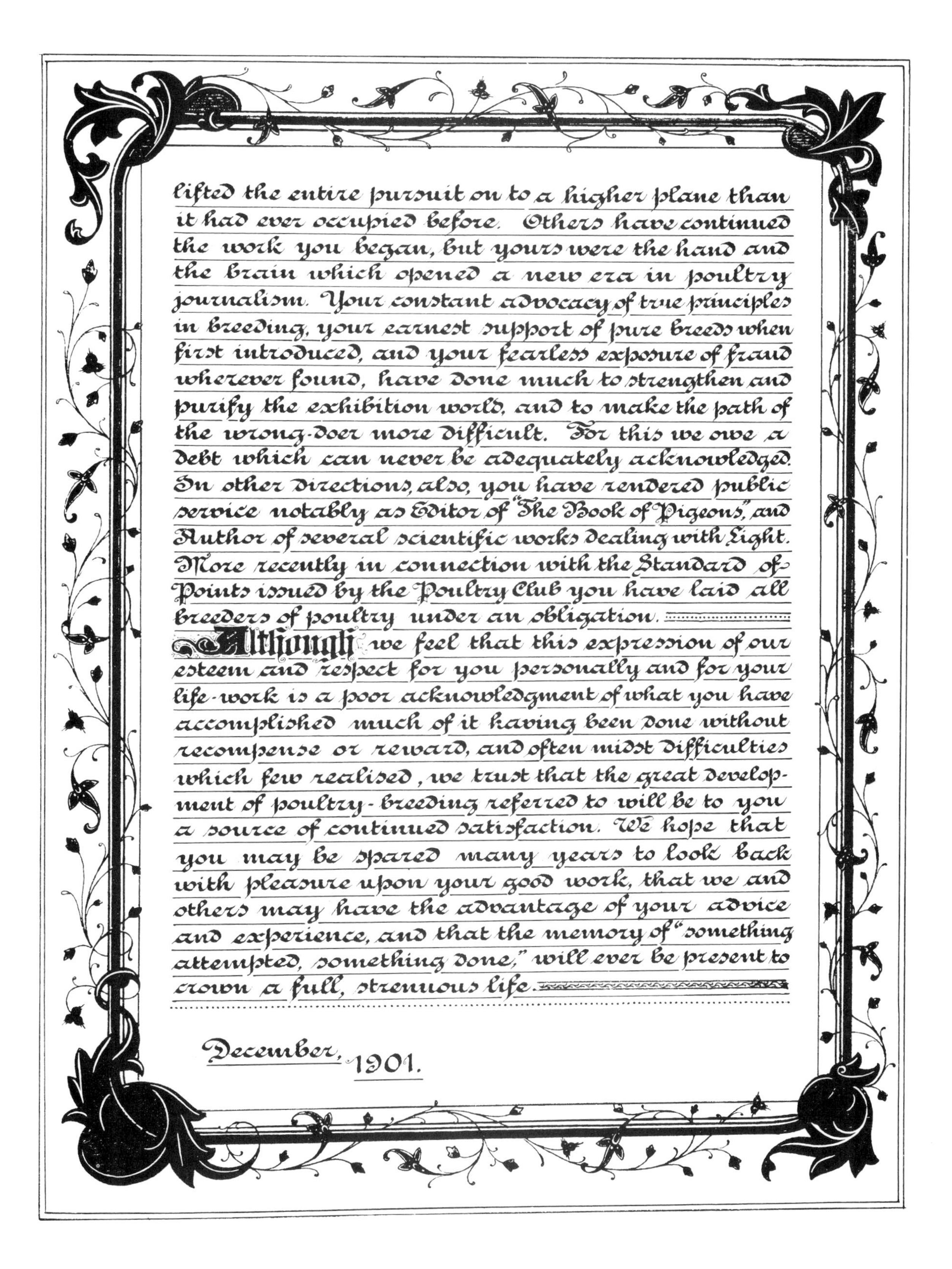

lifted the entire pursuit on to a higher plane than it had ever occupied before. Others have continued the work you began, but yours were the hand and the brain which opened a new era in poultry journalism. Your constant advocacy of true principles in breeding, your earnest support of pure breeds when first introduced, and your fearless exposure of fraud wherever found, have done much to strengthen and purify the exhibition world, and to make the path of the wrong-doer more difficult. For this we owe a debt which can never be adequately acknowledged. In other directions, also, you have rendered public service notably as Editor of "The Book of Pigeons," and Author of several scientific works dealing with Light. More recently in connection with the Standard of Points issued by the Poultry Club you have laid all breeders of poultry under an obligation.

Although we feel that this expression of our esteem and respect for you personally and for your life-work is a poor acknowledgment of what you have accomplished much of it having been done without recompense or reward, and often midst difficulties which few realised, we trust that the great development of poultry-breeding referred to will be to you a source of continued satisfaction. We hope that you may be spared many years to look back with pleasure upon your good work, that we and others may have the advantage of your advice and experience, and that the memory of "something attempted, something done," will ever be present to crown a full, strenuous life.

December, 1901.

Edward Brown (later Sir Edward) said of Lewis Wright that he had known him personally for thirty-one years and was an admirer of his writing before then. He believed that the world would be poorer for his sudden demise. He had "maintained his nobleness of character, his integrity and his firm determination to uphold what he believed to be right and true. His work will live after him through all time".

Today, almost eighty years later, the poultry fancier still knows Lewis Wright through the books he wrote. Not all realise that the development of the *standards* and the absence of any faking or malpractices at shows is largely due to the pioneering efforts of one many who showed the way for others to follow.

1. Comb.
2. Face.
3. Wattles.
4. Deaf-ear or Ear-lobe.
5. Hackle.
6. Breast.
7. Back.
8. Saddle.
9. Saddle-hackles.
10. Sickles.
11. Tail-coverts.
12. True Tail-feathers.
13. Wing-bow.
14. Wing-coverts, forming the "bar."
15. Secondaries, lower ends forming the wing or lower butts.
16. Primaries, or flights, not seen when wing is clipped up.
17. Point of breast-bone.
18. Thighs.
19. Hocks.
20. Legs or Shanks.
21. Spur.
22. Toes or Claws.

Figure 2.1 Points of the Fowl

Figure 2.2 Anconas

CHAPTER 2

THE BREEDS

The first edition plates are those published in 1872 (50 in all) and the balance were produced around 1911. Numerous imprints were published after the 1872 edition so it is difficult to be exact in dating the illustrations.

The descriptions which follow are those used by Lewis Wright. Where the breed has changed little no comments are made, but in other cases, such as with Indian Game and Old English Game, notes are given on the variations which have occurred.

It will be appreciated that Lewis Wright's writing spanned a considerable period and this was carried on by others, especially S.H. Lewer, after his death. The period was one of great change. In the First edition of **The Illustrated Book of Poultry** *poultry shows were a novelty and reached a point never achieved again, when breeds such as Modern Game were bred to a state of "perfection" and virtual extinction.*

The paintings by J.W. Ludlow in the first edition reflect the striving for perfection. He painted birds with all their exhibition requirements exaggerated for all to see the "perfect" bird. They are beautiful to behold and, whilst there is no doubt that all breeds possess great charm and beauty, the paintings surpassed anything ever seen in reality. Nevertheless, they represent an important landmark in showing for the first time prize winning birds with all their virtues in evidence and their defects omitted.

In the second edition many of the birds have a more natural look and have to some extent lost the stylised poses. They come nearer to the poultry to be seen roaming in grassland or scratching in a run. The fact remains that for exhibition purposes the fancier must strive for perfection and hopefully come near to the ideals which Lewis Wright sought to present in writing and in art form.

ANCONAS

This breed, it is generally conceded, was imported into this country from Ancona in Italy, where it has been kept in large numbers by the farmers of that district for its utilitarian properties. Certainly it ranks as one of, if not the very best layers extant. We have frequently heard Anconas decried as layers of small eggs, but as with most breeds, we consider this to be more a matter of strain. The strain that we keep lay eggs which average over 2 ozs. in weight, which we consider quite up to, if not above the average.

As we have said before, they were imported into this country for their exceptional laying qualities, and the reception they met with has scarcely been equalled by that of any breed in recent years. They are indeed very profitable fowls from a utility point of view, as they mature very quickly, pullets very often commencing to lay when about eighteen weeks old. The cockerels are also very precocious youngsters, crowing frequently at five or six weeks old. As table fowls they can scarcely be recommended on account of their smallness, but their flesh is excellent in flavour.

It was not long before they made their appearance in the exhibition pen. About the year 1898 a difference of opinion arose amongst Ancona breeders as to the type of bird which should constitute the standard, and at a meeting held at the Dairy Show in 1899, which was well attended by the principal breeders, a standard was drawn up and passed as a guidance to breeders what to breed for. This standard met with much opposition at the time, and was the subject of much controversy in the poultry papers. The question finally resolved itself into how large or how small the tipping at the end of the feather should be, also the way the feathers should be tipped. We favoured, as we do still, the small V-shaped tip. Were they tipped or mottled – call it what you like – to the extent of three-eighths of an inch, as some breeders contended, on each feather, the bird would present almost the appearance of a white one, as the ground-colour would be covered by the feathers overlapping each other. The Anconas often seen at exhibitions a few years ago were frequently held up to ridicule for their mongrel appearance, and it was with sincere desire to improve this splendid utility fowl that from an artistic or exhibition standpoint, and to breed them more uniform in colour and shape, that the present standard was evolved.

We know of no breed that has made more rapid strides in the time towards attaining that end than the Anconas. When we consider the progress made in breeding since the present standard was made, the success achieved has been remarkable, though certainly there is still much room for improvement. In cocks, a few years ago one scarcely saw anything but white tails, tipped with black, which were certainly not uniform with the body colour. Our aim is to breed them with a good beetle-green ground-colour, with each feather tipped with white, throughout the entire bird.

In mating Anconas to produce exhibition birds, one needs to be very careful in the selection of stock birds. Examine each bird carefully, and discard any that have white under-colour. This is a very common fault, and one that breeders should take pains to exterminate. Another evil to be avoided is lacing, by which we mean a white edging round the feather. Choose those with (as nearly as you have them) the V-shaped tip, with the white clearly distinct from the ground-colour, which gives the birds a slaty appearance. Select for the male bird one with firm erect comb, evenly serrated, serrations deeply cut, face a brilliant red, white face in Anconas being a disqualification and not merely a defect; lobes medium in size, almond shape, and white; body colour as nearly to the exhibition standard as possible; legs deep yellow with black mottling evenly distributed. Be careful about the tail. Examine the bird and see that the feathers are black from the skin; many begin white, then are black in the middle of the feather, then white tip at the end. In hens, again, be careful about selecting those with sound under-colour, and in colour we like them rather on the dark side: we mean darker in appearance than required for exhibition. A frequent fault in Anconas is that they carry their tails too high, squirrel fashion; try by all means to breed this out by selecting only those with low tail carriage.

ANDALUSIANS

The variety known by this name is another which can be highly recommended for its laying qualities. Mr. Leworthy writes of these fowls as follows:

> **"They are excellent table fowls, the cocks weighing about seven pounds and hens five to six pounds each. They are very precocious, feathering fast and kindly, and very prolific indeed as layers. Mine average five eggs per week each, and I find the eggs larger than those of any other fowl, even Spanish not excepted. In fact, taking weight as well as number of eggs to be a criterion, I think them the most productive birds of any I know. One of my first hens commenced laying in January, and up to the end of the year she had laid 220 eggs. I may also mention that their eggs are of an exquisitely delicate flavour.**
>
> **"The comb of the cock resembles that of the undubbed Game fowl, but is rather larger; the hen's comb lies over on one side of the face, as in the Spanish, though many hens even yet are bred with comb erect, as in the original birds. The wattles are in proportion to the comb. The face is red, but ear-lobes pure white, and showing up very distinctly from the face, very much as in the Minorcas. The head should be tapered with as little red skin as possible over the eye.**
>
> **"The cock's neck is long, and hackle rather short; the breast full and round; tail large, and carried very high. The legs are long. The general plumage is a blueish shade or slate-colour, clear all over the ground-colour, laced round the edges with black. The hackle-feathers of the cock are a very good blue for the artificial flies used in trout-fishing. The plumage of both sexes is alike, except the hackle and upper-feathers of the cock, which are many shades darker.**
>
> **"There is also a Pile Andalusian, in which the ground-colour is silver, thinly covered with light blue, which forms the pile. These are very beautiful birds, but are rarely seen."**

We have had some personal experience of Andalusian fowls, a friend in whose poultry-yard we took considerable interest having kept a large stock for some years. She possessed the advantages of a good country run, and from first to last gave a most favourable account of their good qualities in every way. They were very moderate eaters, perfectly hardy, and their eggs never failed. One peculiarity, however, has always struck us, and that was their extreme precocity.

The hens occasionally, though rarely, desire to sit; and when they did so they made very good mothers. The same uncertainty as to colour of the chickens was found with this stock, many coming black, and with upright combs, so that much care was required to keep the yard at all true to the feather; but as a safe, useful, and profitable fowl to keep, they could hardly be surpassed.

Some little latitude is allowed in the colour of Andalusians for exhibition. It may vary from a pale dove-colour to a deep slaty blue, and the lacing may be black, or dark blue, or purple; or in many fine specimens can scarcely be observed at all, the blue ground being almost uniform in tint. The cock's hackle and upper plumage should, however, always be very dark in colour to look well, the rich contrast of colour being required. We have seen this portion of his plumage nearly if not quite black, which looks handsome; but the best colour for beauty, and certainly that which harmonises best with the general type of plumage altogether, is a very deep and lustrous purple. A fine bird thus coloured always graces a pen, and if worthily mated, rarely fails to obtain honourable notice.

BRAHMAS

Lewis Wright had quite strong views on the origin of the Brahma fowl. Other writers had different ideas so a great controversy raged for many years. Mr. G.P. Burnham of the U.S.A., author of **A History of Hen Fever** *claimed that he had been responsible for introducing both Dark and Light Brahmas. Lewis Wright contested this view and stated that Burnham, unable to obtain the pure strain, had produced his own by crossing with other breeds.*

Who is correct in the **detail** *of the controversy cannot now be proven with any certainty. Certainly Lewis Wright produced a great deal of evidence to support his views and readers who wish to study this at length are advised to study* **The Illustrated Book of Poultry.** *Further details are examined in Chapter 1.*

Characteristics of Brahmas

Both the Dark and the Light are supposed to be exactly the same in size, shape, and carriage; but this has not always been the case, and can hardly be said to have been absolutely so at the end of 1900. In the earlier days of the breed the Lights were considerably inferior to the Darks, possibly owing to mixture with Cochin strains; now the Lights as a rule are more broad and massive and fluffy than the Darks, with more of the Cochin type. This should not be so; but the truth really is that the Dark Brahma breeders have made rather a more successful fight for the Brahma model, and it is too true that under many judges a Light Brahma cock of the same model as is still often seen in Darks, would stand little chance. Still, the leading points may be preserved in both colours, and may be described as follows.

The Brahma as a rule looks smaller than a Cochin, but is in reality very often heavier, because the Cochin's loose fluff adds to its apparent bulk. On the other hand the Brahma plumage, though far more abundant than formerly, is still desired to be close-fitting. Cockerels six months old should weigh 8 or 9 lbs, and pullets 7 lbs or more; really pure-bred strains are never remarkable for early weights, though these may be forced; but on the other hand they grow more than most other breeds the second year. Adults sometimes reach great weights. We once knew a cock weigh 18 lbs, but he was a brute; 14 lbs and 15 lbs we have come across several times in exhibition specimens, but 11 to 12 lbs is more usual. The heaviest hen we possessed was 11¼ lbs, but we knew several over 12 lbs. As a rule these enormous birds are deficient in symmetry.

The head of the cock should be small, short, and rather wide over the eyes; not enough to give a cruel or Malay expression, but a sort of peculiar archness or intelligence: the beak also should be thick and short, in harmony with the grouse-head model. Many birds lately have had large coarse heads, or else long and slender face and beak, either of which looks very mean. The comb — the "pea-comb" — resembles three small combs pressed into one, the centre being the highest. This should be small, with the centre ridge straight, and the shape preferred is to rise somewhat from front to centre or beyond, and then decrease a little, with a slight arch. Formerly a very common shape was to rise towards a peak behind, and this is still occasionally seen, but looks very ugly. In some cases we have seen a comb evidently of this type originally, with the rising portion at the back cut off, as shown by the glossy scar: trimming of this kind should not pass unnoticed. The face should be fairly smooth, not too hairy. The ear-lobes in all early Brahmas which we can remember were so long as to fall below the wattles, which ought to be rather small, not long and pendulous; this should be sought therefore as a Brahma point. They should be smooth and bright red and free from feathers.

The neck should be very full in hackle, so that it stands out and makes a sort of junction with the head. It should be long, and well arched, which gives grandeur and nobility of appearance; but the more Cochin-shaped birds generally have rather short necks. The shoulders should be wide and flat, not too much gable-shaped, and the back short, the saddle starting from very little behind the base of the hackle. The saddle is, however, to be very long as well as wide, rising uninterruptedly to the tail, with hackles long and abundant, flowing well over the points of the wings. However massive and thick the bird is otherwise, or even fluffy, the saddle feathers should lie close and hard-looking, and rise more and more with almost a concave profile towards the tail, which rises still more upright so as to form part of the same graceful "sweep" of line from end to end. The tail may be nearly upright, but should not be quite so, and in any case should work in with the contour of the saddle, not stand up "out" of it. However profuse in plumage and thick in shape, this Brahma "sweep" should be well seen, and is faithfully depicted in the coloured illustration, which at the same time should be massive enough for anybody.

The real Brahma tail is itself characteristic, but only seen occasionally. It is sometimes described as if the upper sickles diverged, but this is not correct. It is the upper pair of true tail feathers which curve outwards, like the tail of a black-cock, and the fine sickles curl over between their opened ends. In another type of tail, the whole group of feathers appears to spread out laterally. Neither type is very common now, and if we get a handsome tail, with a good sweep, we may be satisfied. The saddle should on no account get narrower towards the tail.

The breast should be deep, broad, and rather prominent, coming well down between the thighs. These and the legs ought to stand wide apart, and the latter be fairly short; but they certainly may be *too* short, though this does not often happen. They are to be feathered as heavily as possible down the outside of the shanks, and to ends of outer and middle toes, with the feather sticking out well, especially under the hocks, which is the difficult place to get good feather. As a rule the inside of shank, and back toe and inner toe also, are more or less fluffed or feathered too, which is not a beauty. The required leg-feather is practically always accompanied in England by vulture-hocks, for which we have already expressed regret, as usually accompanied by coarser skin and deficiency in breast. But there is a great difference in vulture-hocks, some of them projecting far more than others for the same amount of real feather; and very offensive-looking hocks should certainly be penalised, as against those in which more downward direction or curling in, diminishes their prominence. The fluff should be fairly abundant and well covering the thighs, though neither so full nor so downy as in Cochins. The wings are of medium length, considerably longer than a Cochin wing, and should be tucked up rather tightly. The toes should be well spread apart and straight: a curved toe is rather often seen in this breed, perhaps from its weight.

The hen should present the same small head, which in her case looks particularly arch and coquettish when of the the right model, yet with a sweet and gentle expression also. The body should present the same deep-breasted general outline; and in her case also there is the same characteristic difference between her cushion and that of a Cochin, the cushion of the Brahma rising more and more to the tail, which stands out at the end, instead of drooping as in the other Asiatic breed. The fluff is abundant, but should not be globular, the general appearance being rather square than rounded. The cushion should also grow wider and wider towards the tail, though of late many Dark birds have tended to run off narrow and weedy. The Light pullets, on the direct contrary, have often shown such enormous fluff and cushion as closely to resemble the Cochin type. Some breeders and judges evidently prefer this stamp of bird, which is often of great size; but such a model is foreign to the original Brahma and should be discouraged.

Light Brahmas

We must now deal with the colours separately, and will take the Light Brahmas first. In this breed the real ideal of colour has not varied much, though different faults have appeared at different times. The comb and face have been described above; the rest of the head of the cock is pure white, and the hackle below silvery white, more and more striped towards the bottom of the neck with black. This stripe should be as intense and sharp as possible, and run well up the feather, the edges being sharp and the edge of the hackle white. We have seen magnificent birds on the rather dark side, with a black edging as well as striping to the lower hackles, and are bound to say we thought it looked very handsome, but it is considered a fault, and is also a sign of too much colour in the stock. The most difficult part of the hackle to get good marking is where it comes round to the front of the neck. The hackle cannot be too full and flowing, and should slide like glass over the shoulders and back. The head and neck of a Light Brahma cock, both literally and in fact, form the frontispiece to the entire bird. Red or bay is the best colour for the eyes, but the lighter colour is fully recognised.

The shoulder-coverts and back between them are preferred white on surface, but are black quite as often underneath the hackle, at the part often called the cape. The saddle hackles are by most preferred white, but these also are allowed to be striped, though it should be much more thinly than the neck, and general harmony requires that a striped saddle should accompany a rather dark and brilliantly-coloured neck-hackle: personally we rather prefer the striped saddle. The tail is black, except that the top feathers may be narrowly laced with white. The primaries are black, edged or not with white on the outer web: secondaries, white on outer web, and generally on a little of inside web, rest of the inner web black. The rest of the body and fluff is pure white on the surface, with either white or bluish-grey under-fluff; the shank-feather may be either white, or preferably with some black mottling. The shanks should be brilliant yellow. All the black except in shank-feather, should have as much green gloss as possible.

The hen is the same general colour, the head being white, and the hackle lower down broadly and densely and sharply striped with black. Her tail is black, or may be laced with white, and primaries and secondaries of the wings, and shank-feathers, are as in the cock. The rest of the body pure white all over in surface-colour, except the rearmost tail coverts, which are black edged with white, and the cape between shoulders, which is often more or less black. The under-fluff may be either white or grey as in the cock.

Breeding Light Brahmas

We have now to consider how such birds are to be bred: task which is not easy according to the high standard of the present day. To have any chance of success, a cock or cockerel *must* be selected with neat head and comb: the male bird is all-important in these points, and the day is gone by when a coarse head and a bloated mass of red can win at good shows. The bird must, of course, also be typical in shape and size; but on the other hand, if he is really good in other points, there is no need that he should be large. Size is the one point in which, if necessary, expenditure may be saved.

The next important point to look after is the quality of his colour; by which we mean, not the amount of black he may have, but that his white is clear skim-milk white, and the black a dense black. The same remarks about what is called "sap in feathers" while growing, made in the case of Cochins, apply here also. If such a bird can possibly be got, one that has been white as a cockerel, and moulted white, and grown his feathers white whilst in the quill, will save a lot of trouble in breeding out yellow tinge later. The point is not quite so vital in hens, but in their case also this pure real white, not cream, is the point of greatest value. Brown or indistinct striping in the hackles, will also entail tedious work later on to breed them good.

The *amount* of colour in the birds mated is the final thing to be considered, so far as colour is in question. If a cockerel as first described can be secured of the exhibition type, and also pullets or hens of the *ideal* exhibition type, that is, with hackles really shapely and densely striped and nicely edged tail coverts, and both sexes are really free from all surface splashes or ticks in undesirable places, this mating will usually produce good chickens of both sexes.

Dark Brahmas

The head of a Dark Brahma cock should be almost similar to that described for the Light, but though correctly called white, there is a sort of pearly grey about this white rather different from the snow white of the other. In the neck hackle there is less latitude of colour than in the Light variety, it being essential that the neck hackle should be broadly and densely striped with brilliant black, extending well up the feathers and with any white shaft showing as little as possible, the rest of the feather being a brilliant white, unless occasionally the extreme edge may show a sharp and very fine black edge near the tip. The saddle hackle is also brilliant white, more finely striped, and generally showing rather more white shaft in the feather. What little there is of back should be silvery white, but between the shoulders, where the hackle flows over it, the feathers are black laced round with white: a cockerel has much more black here than a bird in his second year, and black often mixes partly in the white of the back also. The saddle hackles gradually merge at the rear into the tail coverts, which are more and more broadly striped with black, until next to the tail they become quite black, the tail and sickles also being black, unless the upper pair of black-cock feathers are thinly laced with white: white in the tail otherwise is a great blemish, not uncommon in the second year. The shoulders and wing-bows are silvery white in old birds, usually mixed with black in young ones, but should be free from brown or red feathers: the wing-coverts form a brilliant black bar across the wing: the secondaries are white on the outer web except a large spot at the end, this and the inner web being black, leaving the wing very white: primaries black, with or without a white edge on outer web. All the black feathers of thse upper parts should be brilliantly glossed, a green gloss being best, but a purple tinge is not a very serious defect. The breast and thighs and underparts are preferred solid black in an exhibition bird, but small mottling on the breast and lacing on the fluff, or fine white lacing on breast and fluff, are in theory and standard admissible, though practically they seldom win: the shank-feather should mainly correspond with the other underparts, but with even a black breast, a little white there is quite permissable. A real silvery whiteness in the white, free from yellow or straw, is one of the chief points in colour. The shanks should be deep yellow or orange, but any distinct yellow is sufficient.

The colour of the pullets now fashionable is a pure grey ground colour with black or nearly black pencillings, as nearly uniform as possible all over the body, breast, back, wings, fluff, and leg feather, with black tail and striped hackles. This colour used generally to show brown areas or patches, and moult brown in the hens, but recently many birds are practically clear all over, and the best breeders have a fair proportion that moult out as clear. This breeding for colour and marking affected the type somewhat for a time, but we are pleased to see that the general character of the breed is returning without losing any of its beauty.

A dark grey ground also holds its own to some extent, and most of such birds are superior in size to the preceding, and some of them make really fine hens. Some of these have very fair striped hackles, and are usually better layers than the paper-ground birds.

Breeding Dark Brahmas

These changes in colour have very much altered the breeding of Brahmas for exhibition, so that it is now impossible to breed the fashionable colours good in both sexes, from the same breeding pen. This is, however, entirely owing to the change in colour, and it is quite wrong to assert, as some breeders do, that black-breasted cockerels and well-pencilled pullets — so far as pencilling only goes — cannot be bred together successfully, or from the same stock.

To breed exhibition cockerels, the first essential is, as in all cases of double mating, a sufficiently good exhibition cockerel or cock in all points of head, shape, colour, and plumage: the one point which may be perhaps spared to save the pocket, is size, which is not very important if his hens are massive and large. The points need not be repeated in detail, laying emphasis only upon the colour of the white, which is all important. Nothing but vexation can come of breeding from yellow birds, or any with red about the back or shoulders; and while the exhibitor saves the colour of a good bird all he can by keeping him out of the sun, on the other hand, if in his choice for breeding he can pick up one that has been out a great deal in all weathers, *and yet* kept his silvery colour, such a one is twice the value. There are such birds, and there is that difference in strains. This bird we choose on his own merits, and no more need be said.

Of the hens to put with him, among all the mixture of strains one important point is that they be more or less of the same cock-breeding strain: for this we have to trust to personal knowledge or inquiry. The only "points" that can be certainly laid down are, that they must have good heads with small neat combs, and solidly-striped hackles. The body-colour of good cock-breeders varies enormously. We have seen some nearly black, others black pencilled on brown, others covered with minute microscopic pencilling, hardly visible, on a dark slate-colour; very rarely a really well-pencilled blue-grey hen will breed good cockerels, but generally the offspring of such are spotted on the breast. As a rule these hens are pretty dark, and at least very poorly marked; and supposing they have good hackles — that is one point which really can be seen and is essential — the blood of good exhibition birds in their veins is the main thing. If the cock is short of shank-feather, they must of course have abundance, but that is obvious. In starting a new strain, if the cock or cockerel is really good, however poor the first year's produce may be, breeding back to the cock the next season will generally give an adequate reward, and so on while line-breeding is carried on.

The pullet-breeding pen will be entirely different. Here we must select the females from the exhibition point of view, at least as regards colour and markings, and be especially severe as regards breast-pencilling under the very throat. It is an excellent plan to have one or two birds rather darker than the rest in starting a new pen, and to keep watch which hit the best. The great point is uniformity of marking. It is very common to see good breasts with cushion-markings so small as to be indistinct; or there may be a brown patch on the wing. A hen which was clear as a pullet and moulted clear, is especially valuable. If possible select some striped hackles and some pencilled, with a view to avoid getting the whole strain pencilled all up the neck. Most pullets from a clear ground strain, will have what may be called partly-pencilled hackles, which moult out more pencilled still. Pullets of the darker steel-grey type are much more easy to obtain with properly striped hackles. The great faults of the clear pullets at present are want of size, and long narrow backs, and these faults should as far as possible be avoided.

In choosing the male bird, again we must start with the condition that he be of the right strain; because even cock-breeding strains occasionally produce a mottled or laced-breasted cockerel, and such a one will not answer. Supposing that, his chief points are dense striping in his hackles, and all the better if with a hair-line edge. This should be free from white streak as drawn, and the tolerance of white shaft or streak is the chief cause of the prevalence of pencilled hackles in the pullets, which may to some extent be kept in abeyance by choosing solid striping in the cockerels. The cape, or back under the hackles, should be black feathers well laced with white; and the edges of the lacing at the tail coverts should be sharp and clear. The best colour for the cockerel's breast is a narrow lacing with white on each feather; but another marking which answers well, especially with the darker pullets, is a small pear-shaped white spot on the tip of each, with the shaft showing as a black line up its centre. The cockerel or cock should also be compared with his mates in regard to their respective colour fore and aft. If the hens have perfect breasts but poorly-marked cushion, the cock should have narrow but intense stripes in his saddle and tail-coverts, and sharply laced fluff; if the hens or pullets fail near the throat, the male should be narrower in lacing up there, than he is on the fluff, and on no accout have a white cravat.

Some breeders of Brahmas have always objected to the double-mating system. Amongst them is Mrs A. Campbell, of Uley,* in Gloucestershire, who wrote in 1900:

"I have always bred my Dark Brahmas of both sexes from the same pen. I do not at all admire the snaky-headed pullets which now win, which have neither the size, nor the shape, nor the feather of the old-fashioned Brahma. I think it a pity that one shade of colour should be thought of more importance than all these, and live in hopes of the fashion changing to the darker shades; it is such a satisfaction to see one's hens improving, if anything, after each moult.

"A few years ago, at the 'Royal' show held at Chester in 1893, I won first in cockerels and pullets with own brother and sister, and a splendid layer she was; she laid before the show in June, and when she came back she laid seven days running; and in her second year laid about 23 eggs per month in the four months of January, February, March, and April, before getting broody — so show birds are sometimes 'utility' birds too. Then I won the Poultry Club medal with both a cock and hen, own brother and sister, and in 1894 I won at Tunbridge Wells with brother and sister: in the former case the cock won altogether at 26 shows and his sister at 32 shows. I admit that I get a larger percentage of cockerels for show than I do pullets.

"The loss of size from this new craze is I think worst of all. The 'two pen' system I am sure hinders many from taking up the breed for exhibition, for not every one has space enough. I shall stick to the old way, and believe I shall win with it still."

CAMPINES

The fowls known as Campines are undoubtedly of great antiquity; and it is in fact quite evident, now we have them over here, that they exactly answer to the *G. turcica* or Turkish fowl of old Aldrovandus, which has been already alluded to in treating of Pencilled Hamburghs, and which is pictured as a single-combed breed, with pencilling like a Hamburgh's, and the cock's body pencilled like the hen's. Further,

*** Editor's note: presumably the Mrs Campbell who created the Khaki Campbell Ducks by crossing an Indian Runner duck and a Rouen drake.**

there are found both single and rose-combed Campines in Belgium, though single combs are adopted in England to keep the breeds as distinct as possible; and we have said already that single combs still appear in Hamburghs, and that old representations of these also depict pencilled cocks as well as hens. All the evidence is convergent, and points to the breed here mentioned as being the original of the old Chittiprat or Everyday Layer, before the latter had been refined in head and pencilling, at the expense of some loss in size, hardihood, and prolificacy. It is a hardy fowl and prolific layer that the old breedhas been re-introduced into this country.

The plumage of the hens is very similar to our Pencilled Hamburghs, but the pencilling is broader. In cocks the breast and wing-bar are pencilled like the hens, the back and saddle in most specimens being at present white. The main tail is black, and the sickles in the best specimens are edged with white or mackerel markings. The combs are single, erect in the cocks, and falling over in hens like a Leghorn's. After the plumage and smart, graceful carriage, the most striking characteristic is the dark, apparently black, full, prominent eye; when closely observed, however, it is found that the pupil only is black, the iris being a very dark brown.

The chicks are hardy, and feather quickly. I have had them successfully reared on stiff clay, with an excess of moisture for weeks together. In a fairly extensive experience, they are the most precocious youngsters I have ever come across. When hatched in fine weather, I have has the cockerels crowing under five weeks. The colour, when hatched, of both Golds and Silvers is dark brown, the Golds being a shade richer.

The English type of Campine includes both Braekel and Campine of Belgium, ignoring their differences. It might be more accurate to say we are producing a composite bird out of these two, and in America they intend to do the same, modifying our type slightly.

The cockerel has a medium sized, even comb with about five serrations, and the tip slightly following but standing clear of the nap of the neck. It should be fine, even, and upright without a thumb-mark in front, wide enough to be well set, and yet not so beefy as to appear ugly or to interfere with the comfort of the bird. The wattles are medium-sized and of fine texture. The earlobe satin white, smooth, not wrinkled, of moderate size; the beak should be horn colour and the eye black (though the iris is often dark brown). The neck hackle is silvery white, and should be about three times at least as wide as the white ground colour, the latter being wide enough to be seen at a distance, and not a cobweblike marking visible only when handled. The black should be pure black, free from grey and covered with a beautiful green sheen, which makes the bird (of a naturally gay carriage) look resplendent in the sunshine. The idea is that the markings should suggest rings round the body, the geometrical regularity being broken by the rounded white bar at the end of the feather. In the hen the comb may fall over, but should not be so large as to interefere with the bird's comfort. At the base of the comb of the female there is often a bluish zone. This is due to a pigment termed "negresse," and which is responsible for the delicate flavour of the Campine. The colour of the legs is leaden blue with horn toenails.

In England we have only the Silver and Gold Campines, but in Belgium they have some other varieties. The description of the Silver applies to the Gold, substituting gold ground colour for white.

Qualities. — The Campine is primarily a prolific layer of a large white egg, and it is more an "all the year round" layer than one who lays heavily at a certain season. As a table bird it is excellent in quality, and it has the merit of carrying the greater part of its flesh on the breast. It is small in bone, and consequently the ratio of flesh to offal is higher than in any other breed. Judged by weight only, it is not big enough for a table bird; but weight should not be the consideration, as two Campine cockerels weighing 8 lbs. would feed more people and more satisfactorily than an 8 lb. cockerel, as weight means bone, which is expensive to build and useless on the table.

Campine cockerels are useful for crossing, as they improve the number and size of eggs, the quality of the flesh and the amount of meat on the breast. They also improve the hardiness of the progency and the quickness with which they mature for market.

The weight of the egg is not less than 2 ozs., and ranges between 2 ozs. and 2¼ ozs. The colour is pure white and the flavour is delicate. Mr. Edward Brown in his "Report on the Poultry Industry in Belgium," says (on page 38), "The claim is made, however, that eighteen Braekel eggs give equal results in cake-making to twenty-two from any other breed."

Campine are non-sitters. They are excellent foragers and find a great deal of their living on a free range. They are small eaters. They are tame birds, not at all shy or wild, they crowd round the attendant and impede his progress; if frightened they certainly can fly, but don't do so as a rule.

Figure 2.3 Silver Campines

The chickens hatch out strongly and are easily reared and feather quickly. They are very active. Care should be taken not to allow them to get into long grass when it rains, otherwise the mortality is heavy, as they go right into the middle of it to search for food and get wet to the skin, and wet is the worst enemy of all young life. They are precocious. I have seen cockerels four to five weeks old attempting to crow. From the fifth week onwards they present a matured appearance, and those unacquainted with them think they are bantams, so grown up they look. The pullets also mature quickly and begin to lay from four to four and a half months old.

The breed is hardy, but it requires to be well-housed in a well-ventilated house. It does not flourish when the sleeping accommodation is not sufficient. Those that have flourished best with me have roosted in the apple trees.

In type I prefer to see a wedge-shaped bird with a well-rounded breast, the bird presenting a gay and fine appearance and devoid of all coarseness. The tail should be well furnished, the two long sickle feathers standing out at an angle of forty-five degrees from the body. There should also be a full supply of saddle hackle.

Exhibition. — There is no difficulty in exhibiting them, as they are always in condition and rarely want washing. Three days in the pen is sufficient to get them trained for show. Cockerels and pullets can be bred from the same pet. The first cockerel and first pullet at both Dairy and Palace, 1908, were brother and sister. The birds do not look so striking in the show pen as they do on a green run or even in the hand.

The difficulty in breeding exhibition specimens is to combine in one and the same bird (a) good neck hackle, (b) good breast, and (c) good markings on the upper part of the body. It is easy to get any two, but very difficult to get all three.

A great advance has been made in the last four years, but there are still a great many difficulties to overcome, and the Campine offers a fair field to those who love breeding. In a finished breed the novice stands no chance against the experienced breeder, but with the breed in its present state luck may come in and help the novice and put him on equal terms with the most expert breeders; this I have seen happen many a time.

Mating. – Advice on mating must always remain a difficult matter as long as there is more than one way of obtaining the same result. I can only just give a few hints as to how to proceed. First select the male; it is the easiest (if not the only) way to determine first on the head of the pen. See he has the right head points. Foremost see he is pure in colour, black being black and white white, with as much green lustre as possible. Let there be no glaring fault, as red eye, for instance. Find out his weak points, and see the hens are strong in these points. Get well developed hens and get them as regularly marked as possible, and get those that are most free from mossiness and have the best barred tips to their feathers. Working on these lines, the novice will not go far wrong, and in a few years should be able to pick out his breeders instinctively.

COCHINS

ORIGIN OF COCHINS

Books of much pretension have traced the origin of this breed to some fowls imported in 1843, which afterwards became the property of the late Queen Victoria, under the name of Cochin China fowls. As regards the fowls themselves this is a total mistake. A drawing was given in the *Illustrated London News* of that date, from which and the description it is manifest that they had absolutely no points of the Cochin at all, save perhaps yellow legs and large size. The shanks were long and bare, the heads carried back instead of forward, the tail large and carried high, the back long and sloping to the tail, the eyes black, the plumage close and hard. Of what we may call Malay blood they probably had a great deal; of Cochin blood none, or but some trace in a cross. But one thing about them there was: these fowls were not only big, but they probably really did come from Cochin China, and from them and that fact came undoubtedly the *name,* which will now belong, while poultry breeding lasts, to another fowl that has no right to it at all.

The real stock first reached this country in 1847, Mr Moody, in Hampshire, and Mr Alfred Sturgeon, of Gray's, Essex, both receiving stock in that year. Mr Moody's, so far as we can learn, were inferior in character and leg-feather to Mr Sturgeon's, but were very large and of the same broad type; and all alike came from the port of Shanghae or its neighbourhood. The birds were undoubtedly Shanghaes, and had never been near Cochin China; and for years attempts were made to put this matter straight. The first *Poultry Book* of Wingfield and Johnson (1853) wrote of them as Shanghaes, and all American writers strove for the same name years after the attempt had been abandoned in England; but it was no use. We never knew a case yet where facts struggled against a popular name, but the name won in the end, and so it was here. The public had got to know the new big fowls as Cochins, and would use no other word; and so the name stuck, in the teeth of the facts, and holds the field to this day.

Mr Sturgeon's stock, with subsequent imports from Shanghae, has been the main source from which Cochins were bred in this country; America has had many independent importations. Mr Punchard's stock was mainly from Mr Sturgeon, the latter keeping from choice the lemons and buffs, while Mr Punchard had the dark birds which originated the partridges, colour being very uncertain in the breeding of those days.

Characteristics of Cochins

Turning next to the Cochin as it is bred today, its great characteristic, above all, is *massiveness* of appearance, especially in the buffs, which are superior to the other colours, as a rule, in Cochin character. The bird really is very large and heavy, a full-grown cock weighing from 10 lbs to 13 lbs; but a good bird looks larger for his weight than any other breed, owing to the fluffy plumage. This is thinner in the quill, broader in the web, and with more length of loose fibrils from the root than other breeds, thus standing out more from the body, and making it look larger, even on those parts where it appears to lie close. The comb is single and straight, only medium in size, with neat top outline and serration; and these and the wattles and the face and lobes should be smooth and fine in texture, not red-pimpled all over. The head should be small, with a gentle and intelligent look; the neck rather short, and very full towards bottom of the hackle, which flows well over the shoulders. These are wide and flat, and the back so short that the saddle or cushion seems to rise to the stern almost from the base of the hackle. The saddle or cushion must be very broad, and rise well, all but burying the short tail of the hen; the tail of the cock should be as short, and the coverts or sickles as soft as possible, the whole forming a sort of smooth line with the saddle hackles. The body is deep and large every way, the fluff on cushion and thighs standing out as profusely as possible; but the wings are not quite so tight-feathered, or clipped in so close as formerly, but themselves carried more loosely from the body, so that the thigh-fluff in most birds does not show such marked "globes" as it did some years ago. The breast should come down very deep, and be well covered with soft plumage. The shank-feather should be very abundant, and stand well out from the shanks, especially at the weak place just under the hocks. In the hocks, as little of *projecting* stiff quill as possible is preferred, and to be sought for. Vulture-hock is objectionable, but the heavy feather down the shank must not be mistaken for vulture-hock. Close examination would reveal the different nature of the feather, for vulture-hock means hard quilled feathers standing out from the back of the hock. The shanks must be short and set wide apart, and the feather extend to the end of outer and middle toes. The attitude is rather forward, with the stern carried high, and the head (in comparison with most breeds) rather low, and the carriage is dignified.

Buff Cochins

Coming now to the varieties of Cochins, at the head of them all stands the class of colours now all known as *Buffs*. As already observed, at an early date the buff colours were much sub-divided, ranging from the lightest silver-buffs and silver-cinnamons, through lemons and buffs, to the deep-coloured cinnamons, which would now be called almost red. The lightest of these colours were very pretty, the breasts being so pale as to be almost a French grey, while the hackles and top plumage of the cockerels were much darker. The propensity for uniform colour all over displaced these variegated colours, and then for some years the classes were headed "cinnamon and buff". The colour of many birds was still lacking in uniformity, and for several years cockerels occasionally won, which would now be called "tri-coloured", the breast being lemon or orange buff, the hackles and saddle much darker, and the wing darker still, even a red. Such birds did not breed well, besides their variegated appearance, and would not be tolerated in any decent competition. It may be stated broadly that the chief thing now desired is *uniformity* of colour all over in buff Cochins. Of course the hackle, from its different texture, has a somewhat different appearance, and more solid, if not deeper tint than the body colour, but the *tone* of the whole is desired as uniform as possible.

The great essential in breeding self-coloured buffs, is freedom from any "meal" in the buff, or white anywhere in the plumage so far as growing stock is concerned (there may be some come in good old birds, as indicated by Mr Procter above), and good buff under-colour. The latter means that the fluff at the base of the feather is to be buff, not white, or "buff to the skin", as it is termed, and the shaft of the feather also buff, not white. The web of the feather and surface of the bird may be buff, but if the under-colour be white or nearly so, good chickens will be very few. Many birds in any new strain may be very pale buff in the under-fluff at first, almost white; but as selection proceeds darker and darker should be chosen, until fairly rich colour is obtained. After that the breeder will have less trouble.

Mealiness in colour does not mean *mottling*, as when feathers bleach in the sun; it means a most minute speckle of white among the buff, like specks of flour, so fine as almost to need a magnifying glass to see it, and leaving the bird apparently quite a nice even buff, looked at carelessly. Such a bird will never breed good coloured stock; but it is just such as are most generally free from black in tail, and often get chosen on that account; while on the other hand a cockerel with black in his tail, of the proper sound colour, and "buff to the skin", may breed most successfully, and be a very valuable bird.

Partridge Cochins

The following notes are from Mr Richard Southern, of Worsley, well known as one of the oldest and most successful breeders of Partridge Cochins, from the days of the late Mr Edward Tudman downwards:-

"To breed Partridge Cochin cocks is now a very difficult task, the reason of which is I believe that they have got mixed up with the pullet strain, which tends to be brown in the fluff, and too plain hackles. I have not bred the cock strain now for many years, but believe there is always a danger in breeding from birds not black up to the throat and darkly striped to the hackle.

"To breed pullets the first thing is to choose your hens, which should, of course, be the largest and best-shaped possible, short on the leg, and plenty of foot-feather, with, if possible, nice soft hocks; but above all things must be heavily pencilled from head to foot, the breast in particular. I have always bred my very best pullets from hens heavily pencilled in the hackle, and find these always breed the best pencilled ones, so that my strain of Partridge Cochins may be called a pencilled-hackled strain, and I have had it for over twenty years. These hens breed the striped hackle as well.

"The cocks to mate with these hens should have a rich orange-coloured hackle, broadly striped with black, with a few brown *spots* on his breast, but not brown *patches;* he should also be just a little tinged with brown on his fluff, and if his tail feathers have a very narrow edging of brown or bay I like him all the better, as this tends to breed pencilled tails in his pullets. The hens should be chosen a little darker than is required, as they will breed pullets lighter than themselves in regard to the greater portion. I have always chosen my breeding cocks from one to two or three years old, and the hens the same, and never breed from cockerels and pullets. One reason is that I have had many pullets that did not moult out as I should have liked, and it is much safer to breed from those hens which have improved in pencilling up to one or two years old. I also choose my hens and cocks (for breeding) with the shaft of the feather almost black to the skin."

Black Cochins*

Black Cochins have been known from the earliest days, but until comparatively recent years have been little bred in comparison with other varieties, and generally behind them in Cochin quality. For their slow progress in early years there were several reasons; one being the predominant rage for buffs and whites, and another the scarcity of stock, which led to in-breeding, and caused want of size and weediness of build, the proper methods of line-breeding not being then understood. Another reason was the attempt to keep up bright yellow shanks, which all black fowls strongly resist. From one cause and another they had become nearly extinct, when the introduction of Langshans in 1872 gave strong fresh blood, renewed size, and better colour. Such a cross now would be hopeless; but the illustration of the original Langshans on a later page will show that at that date the poor existing Black Cochins had little to lose even in model from the new blood as then known, and the chief change really wrought by it was in the question of leg colour. For a time this was generally quite dark, as in the Langshan, but gradually a very dusky yellow came to be generally recognised as the correct type, and the strengthened stock has been, with the aid of other Cochin crosses, bred up to its present standard.

White Cochins

White Cochins were shown of very high quality from quite early days, those exhibited by Mrs Herbert then, not being surpassed for many years: in regard to hens, indeed, many good judges considered that the Whites surpassed all others in Cochin points and development at this period, though of the cocks so much could not be said, the tails in particular being as a rule too long. One reason for this curious superiority of White hens over other Cochins, probably lay in the fact that much heavier leg-feather was bred for; yet since body shape and feather has been improving, the leg feathering is not so good as in the earlier days. However, the past five years has seen a great advance made in White Cochins, and in point of popularity they run Buffs very close. There is still trouble in getting a pure-coloured cock, though occasionally one is shown with plumage white as the driven snow.

The chief difficulty in breeding White Cochins is of course that of colour; and beyond that, the greatest desideratum at present is probably size. Thirty years ago there was a strain of Whites which had a well-known tendency to show a kind of reddish-sandy colour as a faint stain in the cock's wings, quite distinct from that yellow tinge which is the more common fault; but this sandy strain seems to have died out or been bred out, for we have not seen it lately. The main point is to select birds which were not only white as chickens, but have moulted white, and kept white in moulting. The last test will probably prove too severe for any new strain, the great majority of cockerels showing yellow — what is called "sap" in the feathers — while the latter are growing out of the quills; but as soon as possible this crucial test should be applied, and the very few which do grow the young feathers white throughout, given the preference. When once this stage is reached there will be less trouble in regard to colour; but shade will

*** Wright does not include Blue as a standard colour although this was produced later.**

Figure 2.4 Black Cochins

always be necessary for the male birds as soon as the permanent plumage begins to appear.

In regard to size, the White Cochin does not fall far behind the other colours. By selecting some big hens, with great wealth of feather, there is no reason why Whites should not be produced equal to the Buffs for size, while in shape they are well up to the standard. Much can be done by careful and regular feeding, and anyone wanting size must be prepared for early and late meals with the youngsters. To obtain purity of colour it is essential that the cock should be pure. Many successful breeders have tried a black hen in with the white cock, the idea being to get a dead white plumage; but the silvery feather so necessary will not come the first year, but from the second or third cross. The same result can be reached by sticking to the whites, but all coloured birds must be eliminated.

Other points of colour should receive a word or two. Orange or red eyes are greatly to be preferred in Whites, the pearl eye appearing in this variety to be even more prone to blindness than in the others. It is also rather unusually subject to a stain of white in ear-lobes, which should be carefully avoided: very slight cases may sometimes be cured by frequent friction, or a stimulant to the surface. Bright yellow shanks should also receive attention, as pale shanks are apt in their turn to breed white ones, and the next stage may be that green tint which is fatal. Rich colour in the beak should accompany that in the shanks.

Cuckoo Cochins

Cuckoo Cochins have now and then been shown, but have never met with favour. They were no doubt produced by a mixture of dark and white blood, which sooner or later always produces this colour, with a constant tendency to reproduce the black, or white, or coloured feathers which have been its components. To get rid of these foul feathers requires much care and skill in breeding, and the Cuckoo Cochin has never had sufficient admirers to make the attempt very successful. In fact the colour does not appear to suit the Cochin type very well, and has now become so identified with the Plymouth Rock. Were it ever to become popular, it would have to be bred in the same way as the barred Plymouth Rock in regard to colour, looking after Cochin points as usual. Most of the few we have seen have been deficient in these latter points, and unless the true Cochin character is predominant, then the colour must die a natural death. Though for years past the Cuckoo has been dormant, a pair was shown at one fixture during 1910, and should they have really fine Cochin character, they would probbly win in the class for mixed colours now so common at many shows.

The bodily characteristics of Cochins require some special care in rearing and management. They are above all breeds prone to lay on fat, both externally and internally: hence maize should be carefully avoided for them, and a most careful watch kept up on too great weight, or signs of laziness. When kept in confinement they require even more than other fowls to be regularly and plentifully supplied with green food: if this is not attended to they are peculiarly apt to suffer from liver disease in some form, though in other respects the breed must be classed as hardy. The same ample supply of green food has much to do with the successful rearing of chickens, keeping the system in a healthy growing state, and preventing premature deposit of fat. Over-crowding is perhaps more prejudicial in rearing chickens of this breed than almost any other, and wasters should therefore be picked out early; there will always be sufficient even as regards Cochin character and plumage, independent of faults in colour.

The plumage naturally requires great care to preserve it in good condition, owing to its profuse and soft character, which makes it easily injured. More than in any other breed, pullets intended for show should therefore be separated in good time from the cockerels. So in regard to washing, while we have already said that as a rule it is little matter how or in what direction feathers are rubbed about, a little care not to rub violently to much against the lie of the feather is advisable in the case of Cochins, the feather being so much weaker than in other breeds. To preserve foot-feather, the bird must never be allowed to run in long or stubbly grass, which rapidly wears down the lower plumes. A grass-run for Cochins meant to be exhibited should be mown and kept like a lawn, short and tender. Neither should they be allowed to scratch much amongst long straw, a course which cleans many other fowls so admirably, but which tends to injure heavy foot-feather by friction.

One of the most tiresome difficulties in Cochin breeding is the propensity to "loose" or "slipped" wings, a propensity more common in them than in Brahmas, which also share it, owing to the greater softness of their plumage. The very first birds imported showed this so strongly, that some of the newspapers of the time described such wings as a peculiar formation, enabling the bird to "double-up" and fold its wing in a peculiar manner. It is strongly hereditary, and should, therefore, be sedulously guarded against in breeding stock; if this be done, individual cases can often be cured. But special care should be taken of any unusually promising cockerel, that he be not driven about or flurried, which we are quite certain has often started this blemish at a critical age.

CRÊVE-COEUR

The Crêvecoeur is another of the oldest of the French breeds, having been described by the late Mr Vivian in *The Poultry Book* of 1853. At that time he possessed two varieties, one all black except that the cock's hackles and saddle feathers were often mingled with gold or red; the other a mixture of mottle of black and white. Blues and Whites have also been exhibited at one time or another, as is so usual with all black breeds; but we have not now seen any but black Crêves for many years. The Crêvecoeur much resembles the Houdan in general type of body, but is of more massive make, with heavier fluff and stern. Originally the crest and muffling were heavier, but the Houdan crest has now been bred up to about as much, in all probability by a Crêve cross, as nearly black Houdans with two-horned combs were at one time very prevalent. The Crêve comb consists of two large coral-red horns, meeting at the base like a letter V. Except in this point, its large size, and the heavier build, the Crêve might almost be described as a large black Polish fowl. The heavy stern and rather ample fluff have often disposed us to believe there has been at some time a Cochin cross, and this is to some extent corroborated by the large appetite, for which the Crêve is remarkable among French breeds.

Economically this breed has changed a great deal since its introduction into England. It always laid a very large white egg; but in the early days these were laid rather sparingly, and the birds were found very delicate and subject to roup, and difficult to rear. But somewhere about 1870 a change took place, either from the stock already in England becoming better acclimatised, or from some other and hardier stock being imported. Mr R.B. Wood reported them in 1872 to be nearly as hardy as Houdans, and good layers; and from that date their reputation in both respects has steadily improved, so that Crêves must now be pronounced excellent layers, equally good table-birds, and hardy fowls. It is the more difficult to understand the great diminution for years past in the number of those who keep them, and why specimens are so rarely seen at ordinary shows. They are distinctly profitable fowls, not difficult to breed to exhibition points, and easily kept within bounds.

DOMINIQUES*

This is perhaps the oldest of the distinctive American breeds, being mentioned in the earliest poultry-books as an indigenous and valued variety. In general characteristics it closely resembles a rose-combed Cuckoo Dorking with four toes, or rather the Scotch Grey, but with brilliant yellow legs, which the Americans seem to admire in all their fowls. The comb should be a neat rose, resembling that of the Hamburghs; face, wattles, and deaf-ears red; legs bright yellow; and the plumage what English fanciers call cuckoo-colour, which is composed of a light bluish-grey ground, crossed with bands of a darker grey or blue. The shape is Dorking, and the size medium, averaging perhaps seven to eight pounds in adult cocks and six pounds in hens.

The Dominique is an excellent layer, very hardy, and good for table. It grows fast and feathers quickly, while its plain homespun suit makes it very suitable for countless localities where more showy or "valuable-looking" fowls would be imprudent or out of place. The general appearance will be sufficiently gathered from our plate, the originals of which are birds of the second and third generation bred from imported stock by Mr T.D. Galpin, of Putney Heath. So far as we know, the progenitors of these birds were the second importation which has taken place, the first being a pen sent to the Birmingham Show of 1870, where they were quickly claimed. We have been unable to trace these first specimens.

The Dominique is superior to the Scotch Grey in laying qualities, and to the Cuckoo Dorking in both this point and hardiness as well. The yellow legs are against it as a market fowl; but we have reason to believe the prejudice in favour of white legs is to some extent wearing off even in London poultry

*Now a rare possibly extinct breed in the UK.

markets. However this may be, we have no hesitation in recommending the Dominique as one of the most generally useful "all round" fowls we know. It is to be regretted that in its "native home" it has been of late comparatively neglected, owing to the preference of imported stock of all kinds; and we hear from all our many American correspondents that it would have been far easier to procure first-class specimens ten years ago than now, many of the old strains having been allowed to become crossed and tainted in blood. The birds sent to Mr Galpin gave evident proof of this careless breeding, many of the first generation of chickens being quite black in colour, and several having single combs; but all fully bore out their character as good and useful fowls. We have heard similar accounts as to the uncertain type of many American yards; but a year or two of even moderately careful breeding will speedily banish all such irregularities, this colour being bred with comparative ease. The best mode of breeding it is to select hens of a pleasing medium shade, such as that desired, and to put with them a cock slightly darker, carefully avoiding birds with either red or black feathers, and as far as possible even hackles of a golden-colour; by which means, with a little patience, a strain may soon be formed that breeds a pure blue-grey. The most difficult point is to get enough depth of colour in the cock without obtaining sickles or tail-coverts nearly or quite *black;* hence it is necessary at first to raise as many chickens as possible from the first stock, in order to have plenty of material for selection. Besides the colour, the neat rose-comb, red ear-lobes, and brilliant yellow legs, with fine shape and size, are all that need to be attended to in breeding Dominiques.

DORKINGS

With the much greater knowledge and experience of poultry which has been accumulated during half a century since exhibitions have been held, it has become more and more certain that the English Dorking, at least, is one breed which we unmistakeably owe to the Roman conquest of Britain. It has been already intimated, on Cæesar's authority, that the ancient Britons did not eat fowls at that date; but the Romans did, and had learnt to select their table fowls with some care; and the Roman writer Columella describes as the best and most esteemed, a bird with all the essential marks of the Dorking race, which there can be no reasonable doubt that the conquerors carried with them into Britain, unless they already found it there, which is scarcely likely.

The Modern Dorking

The Dark or, as it was formerly called, the Coloured Dorking until quite recent years well maintained its position as an exhibition fowl, and still, with the other colours, yields to none in its splendid utility qualities. For the time it has been elbowed aside by newer varieties, some of which owe not a little to the Dorking themselves; but, in common with all who remember the splendid classes of Dorkings in the past at our classic shows, we trust that they may yet enjoy their own again.

The colours most often seen in Dark Dorking hens show nearly black plumage on the back, with the wing-feathers bordered with black round a greyish centre covered with marking, as described in the Standard. At one time birds were shown darker on the wings and cushion than even this; but that extreme has been discarded.

Red Dorkings

One of the very oldest varieties of colour Dorkings, but almost unknown out of the south-eastern district of England, is the Red Dorking. This is a perfectly pure race, never amalgamated with the Dark as the old greys and speckles have become, and in our own opinion entirely free from any cross with the White. They are not so large as the Darks, are very small boned for the size of the body, and the single upright comb is much smaller, thinner, and finer looking than those usual in the Dark or Silver-grey breeds. As table fowls there can be none better; and we cannot but think that if the few who possess them would systematically exhibit them in such Dorking classes as are offered for "Any Other Colour", they would gain more support. The cock is a black-breasted red; the hen more of a brown-red, some of them laced. Taking the colour into account, in connection with the smaller comb, it can hardly fail to be noticed that this variety comes the nearest to the old description of the "best" Roman birds, by

Columella, already referred to. There is perhaps just a chance that some of the colour may be due to a far-back cross with the black-breasted Red Game once so diffused over England; but if such ever did happen, it must have been long ago indeed.

Silver-grey Dorkings

Another beautiful and well-marked variety of Dorking is that known as Silver-grey, which has separate classes at nearly all important shows. In great measure these birds were an offshoot of the preceding, at a time when the Coloured Dorking was often really a **Grey** breed, from which a lighter grey could be selected. But in addition to this process of selection, it is recorded in several quarters that the Silver-greys thus selected from the Dorkings were also crossed with Silver Duckwing Game of Lord Hill's breed, in order to fix the beautiful colour of the hens; and the effects of this cross were to be seen in a slimmer build and different carriage of the shoulders. Another result of the Duckwing cross was a somewhat too silvery colour, leading to a great tendency to splashes of white on the cock's breast and thighs. Both these defects have long since been overcome, and Silver-greys are often now amongst the most typical of Dorkings, and very nearly if not quite equal in weight to the Dark variety.

The head of the Silver-grey cock should be silvery white, the neck-hackle of the same colour, perfectly free from any tinge of straw-colour, but may be (and generally is) streaked with grey in the lower feathers falling on the shoulders. The back and saddle silvery white, shoulders and wing-bow also clear white; wing-bar green-black; wing-bay white with a black upper edge; breast and under-parts and thighs jet-black, without any mottling or grizzling, except that a little on the thighs is tolerated in old birds; tail glossy black, with sound broad sickles. The white parts should have no tinge of straw, and there should be no signs of brown or chestnut bordering the wing-bar or other margins of the black plumage.

The hen's hackle is also silvery white on the head, but lower it becomes striped with black, often with a little longitudinal pencilling. The breast is a rich robin-red or salmon-red, shading off ashy colour on the thighs. The body and wings are a silvery grey ground colour, minutely pencilled over with dark grey, free from black splashes or reddish tinge, and each feather showing the white shaft, but not obtrusively: tail rather darker. The general effect varies in different birds from a bright silver-grey to a softer duller grey, but in any case should be grey. The silvery greys have usually lighter salmon breasts, and are specially apt to breed cockerels with white spots or grizzling on the breast.

White Dorkings

The White Dorking, as already intimated, was in all probability the purest representative of the original race, unless we except the Red, at the time when Bonington Moubray wrote in 1815. It alone — again perhaps excepting the Red, which was never mentioned in those days — from the first always bred the fifth toe, and there are observable certain differences in carriage, and a greater elegance of form, which is usually rather lighter, and without that massive heaviness which probably came largely from the Surrey fowl. There is little doubt that at one time the breed also received a little crossing with Game, and it is curious that some strains to this day lay eggs of a delicate pinky or French white shade; but certainly no cross has taken its place for many years.

In this variety alone, a rose-comb is essential. This should stand up well, with a long and straight leader behind, and be of good shape generally. This is perhaps the most difficult exhibition point, as a large breed like the Dorking requires pushing on somewhat, while on the other hand any meat forcing tends to excess or deformity in the comb. Free range gives the best all-round results; and on the whole, taking comb, and size, and plumage into consideration, there is no variety in which ample range is more important to a really high exhibition standard.

In breeding White Dorkings the chief points to keep in view are good combs and silvery-white colour.

Cuckoo Dorkings

There is yet one more recognised variety, known as the Cuckoo Dorking, this being the old English word for the blue barred plumage called by Americans Dominique, and seen to its greatest perfection in the Plymouth Rock. Cuckoo Dorkings have scarcely been known out of Surrey, and clearly originated in

Figure 2.5 White Dorkings

the crossing for table poultry of dark and white varieties, the colour always appearing when much crossing of that kind takes place. The birds, as we have seen them, have always been somewhat small, but are generally reported as hardier than the more orthodox types under ordinary conditions, and are said also to be more juicy in flesh.

FAVEROLLES

One of the most popular French breeds at present — the Faverolles* — is of later creation. Although the outcome of rather complicated crossing, it quickly took strong hold upon English breeders, owing to its superb table qualities combined with hardiness and quick growth, and was not long in securing the support of a Club and a place in the Standard. The following article upon the points, qualities, and breeding of the Faverolles was contributed by the late Mr J.P.W. Marx, Nottingham, well known as having taken a leading part in popularising and standardising the fowl.

"Faverolles have for some time been common in the northern part of France, where they were regarded as simply useful fowls. They are the result of crosses to produce good layers, particularly in winter, whose chickens are strong, hardy, and quick-growing, with thin, white skin and fine bone, abundantly covered with meat, and lending themselves readily, if need be, to artificial fattening. Brahmas or Cochins, Dorkings, and Houdans were used to produce Faverolles; and as the different varieties of those breeds were used indiscriminately, the Faverolles are met with various colours, yet with well-defined characteristics of habit, shape, and quality. The salmon, and the white or ermine varieties, gradually became most numerous on account of their better laying and table qualities. A few seem to have been kept in England about 1892 or 1893, but little was heard of them till 1896; since then they have become scattered all over the country.

"Whatever the colour of the Faverolles, the general characteristics are the same. In both sexes the comb is single, upright, medium in size, with neat serrations and free from coarseness. This is a difficult point, since of the breeds which were selected to make up the Faverolles, the Dorking alone has a single comb which falls over in the hen. The peculiar combs of the Brahma and Houdan are strongly hereditary, and thus all kinds of combs crop up in the Faverolles, and most careful selection is required to get and retain the correct type. The beard and muffling should be very abundant, the beard thick and full rather than long and thin. These, again, being only found in one of the original breeds — the Houdan — are difficult to breed; indeed, the head of the Faverolles is one of its most characteristic and important features. The head itself is broad and short, with small, thin wattles and stout, short beak. The head should be free from crest, which is nearly bred out; still there remain traces, particularly in the cocks, in the shape of a few short, upright feathers either side of the comb, which would only be noticed by a breeder who has had experience in eradicating crested blood. The short, stout neck is thickly covered with rather close-fitting hackles. The body is broad, deep, and wide; the back very broad and flat; the breast is also broad, with the keel-bone deep and prominent; the whole giving a sturdy, massive look to the fowl. Greater length of keel and back is seen in the hen. The wings show boldly in front, yet are distinctly small. The thighs are short and set wide apart, with the knees quite straight. The shanks are of medium length. A dumpy, short-legged fowl is not wanted, and the excessive shortness of leg detracted very much from an otherwise capital hen which was most successfully shown in 1900. The leg should be fairly stout in bone without being coarse, and be slightly feathered on the outside down to the end of the outer toe. The leg feather should be soft in texture, with no sign of the vulture-hock too frequently met with. The toes are five in number, and the extra or fifth toe, as in the Dorking, should be clear and distinct. The tail feathers and sickles are full and broad; the sickles incline, however, to be short in length, and are carried rather upright, as in the Brahma; a large tail with long sickles carried low or straight is not in keeping with the build of the bird. The tail of the hen is fan-shaped, and carried rather high.

"Cocks should weight 7 lbs to 8½ lbs; hens, 6 lbs to 7 lbs; cockerels, 6½ lbs to 7½ lbs; and pullets, 5 lbs to 6½ lbs. These weights are not excessive, and are often exceeded, though generally at the expense of quality.

"The colour of the Salmon Favorelles cock is quite different from that of the hen. Some are a mixture of black and silvery white, like the Silver Dorking; others, which have the preference, are warmer in colour, like the dark Dorking. In the exhibition salmon cock the beak, legs, and feet are white; any pink colour on the leg should be dealt with severely if it is too prevalent, and should be eradicated. The skin also is white and very fine; a coarse, red skin is a distinct fault. The face, lobes, and wattles are red, nearly concealed by the muffling and beard, which is black, not ticked with white. Neck and saddle-hackles are straw colour, quite free from any stripe, although many cocks still retain the Brahma hackle, and probably will do so for some time yet. The breast is black; very few are sound in breast colour; the majority show white mottling, particularly towards the bottom, others even have feathers tipped with bronze or red. More latitude is allowed with the back and shoulders, which may be a mixture of black, white, and brown. The wing-bow is straw colour, the wing-bar black, and the outside of the secondaries white. The tail, under colour, and thighs are black; the tail coverts may be brown. Some cocks with much less black in them have the breast mottled with red and white, and the back and shoulders a rich red brown; these are very handsome, but not in accord with the present standard.

"The Salmon hen is much like a Wheaten Game. The head and neck are a wheaten brown, broadly striped with a darker brown. Beard and muffling (both are much heavier than in the cock) are a creamy white. Back, shoulders, and wings wheaten brown, the colour running lighter on the sides until it meets the cream colour of the breast, thighs, and under-colour. Primaries, secondaries, and tail are wheaten brown; these at present are very imperfect, for a great deal of black or white, or both, is to be found in most hens. Face, wattles, legs, and feet are the same as in the cocks. The definition of the colour as 'wheaten brown' is not a happy one; it may mean the warm brown of red wheat or the much lighter shade of white wheat, and the latter seems to be the same colour which is required. The fashionable Salmon hen is a warm cream colour with a pale brown colour on her neck, back, and tail; a delicate pink or salmon shade in these colours is preferable to a faded, washed-out white colour. Any trace of buff, gold, or hard brassy colour should be discarded.**

"There is a very handsome strain of what may be called red wheaten brown hens; the back and sides are blotched with a deep chestnut brown, which runs on to the tail, and the hackles are broadly striped with the same colour; they have a rough, hardy look, but are too dark and red for the colouring of the standard.

"The Ermine or Light Faverolles are marked like Light Brahmas, and, remembering their origin, it will be found quite as difficult to obtain the clear, densely striped hackles with pure white body colour free from ticking. The suggestion before given for breeding light Brahmas should be closely followed in mating.

"In mating Salmon Faverolles, comb, width of back and between the thighs should be attended to in both sexes. The comb should be free from side sprigs, and, if possible, of fine quality in the hen, and upright. The best combs procurable should be used, for faults here are sure to appear in the chickens. A cock with heavy beard and muffling is valuable as a breeder. His neck and saddle hackles should be a yellow straw shade in preference to white for cockerel breeding; a slight stripe or ticking of brown or brownish grey may be tolerated in a pullet breeder. Hens with any black in the hackle, even at the tip, should be cautiously bred from, unless it is known that their mother was better than they in hackle colour. The feather itself should be rather short, but broad, to give room for the darker centre. The breast of the cock should be a solid black from throat to thigh; many are ticked with white, and a few have a mottling of red or brown, and these are likely to breed better coloured chickens than those ticked with white. The sounder the black of the thigh and under-colour the better; cocks showing much white, breed cockerels lighter than themselves, and pullets too weak, almost white in under-colour. The tail coverts should be a dark chestnut-brown in a pullet-breeding cock, and the rest of the tail black. The sheen on the black throughout the cock should be a rich metallic bronze, not a beetle green shade. The hens should be as near the standard colour as can be obtained; the weak points are wings and tail, where black and white are sure to be found. Hens with much white in wing should be mated with a bird sound in wing, with very little white ticking in his under-colour. The brown colour of the tail may be improved by selecting a cock with abundance of coppery brown lustre and brown tail coverts; if the tails of his daughters show an improvement, he may be mated up next year with the best of them in that respect. The shaft and down of the feather quite to the skin should be a creamy or wheaten brown; hens with black or ashen grey down throw a number of pullets with black in wing and tail."

*** Favorelles is the name of a place, and should properly be always spelt with a terminal s. Such a word as Faverolle is a barbarism; but it seems creeping in as the English form, and perhaps cannot be helped. The effort should however be made, to which end we make this direct mention of the matter.**

**** Editorial note: Wheatens at this time were much darker than at present.**

FRIZZLED FOWLS

These fowls are properly called by the name here given, which graphically describes the appearance of the plumage; the name of "Friesland" fowls which is often applied to them having no foundation whatever beyond the riduculous attempt to put the proper nomenclature into a more "genteel" form. If any local name at all be given, that suggested by Mr Tollemache would have decidedly the best claim, though the Frizzled fowl is found in nearly every part of the world.

In the course of many years experience in breeding, and intercourse with other amateurs, we have frequently met with birds, of all the Asiatic races especially, in which the neck-hackles had a tendency to twist out of the true sweeping line, towards the back of the neck; a defect well known among Cochin breeders, and less so among Brahma fanciers, as a "twisted hackle". The causes of this defect have usually appeared to us to consist chiefly in either breeding from young birds on both sides, or the exclusive use of too dry a food, such as oatmeal unmixed. However this may be, the fault has a strong disposition to be hereditary; and the tendency to it is very plainly the same which, exaggerated and developed, produces the Frizzled fowl. Indeed, by selecting specimens with such twisted hackles, and breeding them together, birds partially frizzled would almost certainly be produced.

Frizzled fowls occasionally vary in other characteristics, though usually presenting neat rose-combs and short dark legs. The peculiarity is in the plumage; every feather being curled back in the wrong direction, as if the bird had been roughly stroked the wrong way, and presenting a most grotesque appearance. The tail-feathers are not, of course, thus re-curved, but the webs are loose and disconnected. The most usual colour shown in this country is white, but we have seen very handsome brown or rather partridge-coloured specimens, and also black. The last colour is to our fancy the handsomest of all, and we have accordingly selected it for illustration in the plate, which renders the birds to the life; all previous illustrations of this breed which we have seen being the merest caricatures.

Temminck states that the Frizzled fowl is found throughout Southern Asia, Java, Sumatra, and the Phillipines. It is also common in Ceylon (where it is said, however, to have been brought from Batavia), and we have heard of it in the West Indies, to which it is scarcely likely to have been exported. In some old descriptions it is evidently confounded with the Silky fowl, with which it has, however, no connection whatever, the two breeds being quite different and distinct, as may be seen at a glance by comparing the two plates.

We have heard Frizzled fowls called ugly, but cannot ourselves join in such a verdict. We have seen ugly specimens certainly, not only frizzled, but "ragged", which the fowl should by no means be. Birds on which every feather is properly and neatly curled, however, though singular enough in appearance, have a beauty of their own, which will we think be admitted even on the strength of Mr Ludlow's very faithful representation.

In breeding these fowls perfection and neatness in the frizzled plumage must of course be the chief point in choosing stock birds, colour being preserved or modified in the ordinary way.

LA FLÈCHE

The La Flèche fowl has never been so popular in England as either of the preceding, chiefly owing to first importations being somewhat delicate, a defect since remedied; partly also, perhaps, because there was not sufficient of distinctness in its characteristics. The Polish blood in its veins is shown by the two-horned comb, the small crest in many specimens, and the nostrils; but the carriage, and white ear-lobes, and green-black plumage, evidently show a cross with the Spanish fowl, as another progenitor. This breed formerly supplied many of the best fowls for the Paris markets; but both it and the Houdan and Crêve have lately been greatly displaced by more recent creations of the French breeders.

In general appearance the La Flèche is a tall, Spanish-looking fowl, though not now bred quite so tall and upright as formerly. The size is larger than Spanish, fatted specimens often reaching 11 lbs, and usual weights being 8 lbs for cocks and 6 lbs for hens when shown alive. The bird has little apparent Polish "character" about it, being high on leg, with rather long back and flowing tail. The head is smooth-faced and rather long, the comb consisting of two rather small horns of the V character, but standing nearly upright, behind which is often a very small crest of a few short feathers slanting backward; but specimens are preferred without any, and much crest is penalised. The wattles are long and pendulous, the ear-lobes round, of medium size, smooth and pure white. The back should slope somewhat towards the tail, and the keel should be long and straight. The plumage is close, of a glossy green-black, the legs dark slate to almost black. In breeding, the combs, absence or almost absence of crest, and good ear-lobes chiefly need attention, the plumage and general shape being usually very uniform.

This fowl lays large white eggs, which are produced pretty freely and tolerably early under favourable circumstances. On dry, suitable soils La Flèche chickens fledge rapidly, and, with good and nourishing food and freedom from damp, are quite hardy.

LAKENVELDERS

Since Lakenvelders made their appearance in England in 1901, being exhibited for the first time at Shrewsbury Show in June of 1902, there has been a marked improvement of the breed; and all fanciers visiting the Dutch and Belgian shows state that we have better exhibition birds in England than any shown abroad. The reason for this is not hard to find. The Lakenvelder Club may not be a very strong society, but it has done one good thing by issuing a standard, and the members all striving to breed up to it.

Lakenvelders are by no means a new variety, for they can be traced back to the first half of the last century in Germany. Most of the birds imported, however, came from Holland, and we in this country follow the Dutch standard of a black saddle hackle in the cocks, whereas the Germans favoured a white hackle, but a striped or tickled hackle not disqualifying. The hens when first imported were white, with grey necks and tails, and little else can be said for them, as I have not found any reliable Continental standard for the females, and so hens for the first year or so in this country won simply through the pureness of the white. Now we strive for a neck-hackle as dense black as possible, and a solid black tail, with the rest of the plumage a pure white. A hen of this type is really a pretty bird. Not one hen in fifty that came over, however, possessed a black tail, and consequently it is really harder to find a good hen than a good cock.

The main points to breed for in the cocks are a black neck-hackle, with no grey feathers about the head; a black saddle-hackle, and a good black tail, nicely carried, with the rest of the plumage a pure white. The under-colour is more or less grey in both sexes.

Figure 2.6 La Fléche

Figure 2.7 Lakenvelders

MODERN GAME FOWL

DEVELOPMENT OF THE EXHIBITION GAME FOWL

But as Game fowls began to be shown more and more by persons who never fought them nor dreamed of ever doing so, change inevitably crept in, from causes fully explained in the earlier chapters of this work. Judging, as well as breeding, no longer remained in the hands of the old cockers, and details of mere appearance began to be more studied, both in regards to colour and form. In regard to both points, change at first was very moderate, and it crept in gradually, by insensible degrees. Exhibitors and judges understood that the Game fowl was different somehow from the breeds which were often termed in comparison the "heavy cart horse style"; the tendency was natural to prefer somewhat the taller and more reachy birds; and to a certain extent the modified type did, especially while confined to the earlier and more moderate degree, appeal to even the general public with a beauty of its own, and was welcomed by many for the very reason that it *was* somewhat distinct from the original cock-fighting model.

But changes of this kind, when once fairly initiated, could not stop at the point reached when Mr Douglas wrote the above in 1872. So as soon as fanciers and judges began to look specially for height and reach and colour, it was inevitable that they should seek to get more of these points; and they did so. The process and the gradual development by it of the present Exhibition Game fowl are so interesting and instructive, that we have asked Mr Ludlow to prepare in illustration of it the three sketches shown in Fig. 2-8. The centre figure represents the bird in its transition stage, at the time when the above sentences were written, the outline being an exact reproduction on a reduced scale of Mr Douglas's Black-breasted

Red cock, "The Earl", winner of the cup at the Crystal Palace in 1870, which formed the frontispiece to the first edition of *The Illustrated Book of Poultry*. This bird was painted in oil by another artist, as well as by Mr Ludlow; and having seen and compared both paintings, we select it as an absolutely authentic contempory record of the time. The other two outlines are sketched from Old English Game, as representing the original type, and from an exhibition bird of the present day.

It will at once be seen that the Game fowl of 1870, as here reproduced, was in a transitional state of development in regard to other points than height or reach. The powerful "boxing" head and beak were already becoming longer and thinner, though not so long and thin as they have become since. The tail has become much closer and more whipped together, though not nearly so much so as afterwards, and still possessing a singular beauty of proportion of its own, especially in regard to the nicely "Venetianed" arrangement of the sickles and side-feathers one over the other. The neck has become more slender, and the hackles shorter and more scanty, though both these changes also are carried much further in the bird of 1900. It is needless to go into more detail, but the diagrams themselves will form an instructive and impressive object-lesson concerning the profound changes which fashion and breeding and judging can effect in one of the oldest breeds of poultry.

The later stages of this transformation were not effected by selection alone. Even in 1872 the Rev. A.G. Brooke, in writing upon Malays, stated the fact of that breed being used as a cross to increase the size and stature of Game for exhibition, and that applications were made to him for birds to serve that purpose. But subsequently the Malay cross was used quite extensively by breeders in order to attain greater length of limb, with shorter and harder and more scanty feather. With this cross came at first, of course, very bad heads; but these were very soon bred out, as were other prominent Malay points: there has remained, however, a more sloping back as a rule, and more prominent shoulder-butts than belonged to the old English fowl: also a length of limb that compares with, if it does not even exceed, that of the Malay itself.

Dubbing Game

The necessary process of dubbing is best performed with the bird in the hands of an assistant, who understands the proper way of handling a Game fowl. Using a pair of surgical scissors, the operator cuts first from the back of the wattles, taking that part between the blades and cutting towards the beak, being careful not to cut too deeply, or the jaw-bone might be injured. Now that heads are desired so fine, this is best done at six months old, leaving the comb till nearer eight months old. To take off the comb, the operator stands in front of the bird, cutting from the beak to the back of the skull, and keeping the

Figure 2.8 Development of Modern Game

scissors firmly down the head. If the operation is carefully done the wound will heal in a few days, but care must be taken to keep the birds from fighting, as a few minutes' fight before healing has taken place, might probably cause disfigurement for life. It is no use dubbing cockerels before runs can be found for them, as those which have agreed before will always fight after it: the operation has so changed their appearance, that they meet as strangers, and will no longer agree.

Long training in pens is not good for Game. The best way of training for exhibition is to place a bird in a show-pen for two or three days only at a time, with intervals of three or four days between each time, training them to take food from the hand. Then they should be accustomed to feed from the hand held high up in front of the pen, so as to induce them to stand up, come well to the front, and show off well. In general the bird should be made as tame as possible, when he will not mind being handled, and birds once thoroughly trained never seem to forget it.

It is usual to trim the heads of Game cocks a little before exhibition, removing with scissors the line of little spiky feathers at the sides of the amputated comb, close to the head, and also the little feathers which project from the face. Some draw out the latter, as is done in Spanish faces; but in this case they grow again.

Very few words will be sufficient for the controversy which now and then arises about the operation of dubbing Game. Almost without exception the assumed humanitarians have been totally ignorant of Game fowls; and it is gravely to be regretted that the names of some persons in high position should, by the mistaken representations of people of this kind, have been dragged into a crusade of which they also have not been competent to judge, since they have had no knowledge of the facts and conditions. All these good people forget that dubbing originated when shows were unknown, as a practical necessity, and merely because the fighting cock, when undubbed, was fatally handicapped in his battle, and suffered continually during his life. The old cockers simply found that a Game cock was saved a far greater amount of suffering, and often death, by being dubbed; and they dubbed him for this and no other reason: of exhibition they knew nothing at all. Some oral and written evidence that has been quoted from "veterinarians", professing to "prove" that the comb of a fowl, "owing to its profuse supply of nerves, is specially sensitive", has on the contrary proved an ignorance on their own part quite extraordinary. Examination under the microscope of sections from a cock's comb, leads to an exactly opposite conclusion; not, of course, that there are *no* nerves of sensation in the comb, but certainly to the effect that the comb is anything but a specially sensitive part of the body. The same conclusion is suggested by phenomena familiar at times to almost every poultry-keeper with small runs, who will have seen a cock standing while the hens pecked his comb into a miserable state, with apparently entire unconcern. Dubbing was not adopted to avoid injury to the comb, but to leave no hold; because a Game cock strikes with his spur close to where he holds by his beak, and thus the face might be terribly cut, or the eyes torn out, if the comb was left on. We fear there is a large class of people whose notions of cruelty or humanity depend not so much upon *real pain or suffering* — a matter always to be taken into grave consideration — as upon the presence or absence of visible wound, or of a few drops of blood. At all events, we knew of one lady who used very strong language indeed about dubbing Game, who regularly sent a male kitten to be "made into a house cat", for the merest reasons of her own personal convenience; and that sort of thing furnishes food for reflection. Though we have had some share in the active prevention of animal suffering, we never were able to get up strong feeling about *any* operation that only takes a minute or two, has no pain of anticipation, and is apparently forgotten as soon as it is over; and we have repeatedly seen a Game cock begin to feed as soon as tossed down upon the ground. As, however, we believe that some people have been deterred from keeping Game Bantams especially, from dreading the supposed cruelty of this operation, it seems worth while to point out that even a dentist's nitrous oxide "gas" will cause anæsthesia quite long enough to dub a bird; or if that be inconvenient, that a few whiffs of chloroform in a handkerchief will equally prevent any pain whatever.

It is not much use attempting to breed exhibition Game, any more than the Old English, without plenty of room. The cockerels may be kept together until grown, on a good run, under an old cock, as in the case of the older breed; but there is "Game" in them still, however modified, and as fast as they are dubbed they must be provided for separately. Moreover, space is required to rear the chickens of what is now a rather delicate breed, in health and condition. Finally, without range the birds cannot be shown in the hard feather which is so necessary to success. A few split peas daily for two or three weeks before a show will help this to some extent.

In judging Game, style and make and condition are taken into consideration to a greater extent than in any other breed except the preceding, or in the corresponding breeds of Bantams. Hence, keen as is the competition in colour and marking now, if it not infrequently happens that a bird somewhat inferior in these will pull off the honours by great superiority in style and character of plumage: for instance, a somewhat too dark cockerel, or a too rich or a slightly foxy pullet, may be so far superior in make or feather as to win over better colour. The prize birds, or those likely to be chosen at least, should always be handled, the handling counting for a great deal in Game. Handling is also the only way to detect crooked breasts, which of late have crept into this breed more than formerly, so that occasionally, when the apparently best bird in a class is left out, inquiry will elicit the fact that it was on account of a crooked breast.

OLD ENGLISH GAME FOWL

The bird known under this name stands by himself alone. In lineage none may compare with him, since his origin is absolutely lost in sheer antiquity, and when we do first hear of him, he is already of noble blood amongst other fowls. He has for generations been known as "the English fowl" — Buffon writes of him as such; and he has stamped his very name upon our speech, so that when we want to express a dogged courage that does not know how to yield, no matter what hopeless odds there are arrayed opposite, we say that our soldier heroes stood "game" to the last against their foes. He has earned the distinction well, lifting the name out of the very gutter — for it was first given him as being identified with "sport" or "gaming" in the old sense, so that household bills of James I contain entries for the expenses of "cocks of the game for his Highnesse's recreation" — as he fought for his owners with the courage of his race; until at last the higher meaning of the word came not from them who had bestowed it, but from the bird who fought so undauntedly for a meaner master's stakes.

Let none suppose that all summed up in this was unmitigated evil. Rude times require rude virtues, and it will not be forgotten that the original of the very word virtue itself, stood equally for virtue and for courage amongst the Romans. Thus it occurred naturally, that nearly all primitive nations and civilisations deliberately sought to learn from the stubborn valour of the fighting cock. Every schoolboy will remember how Themistocles revived the courage of his soldiers by an example before their eyes of two cocks fighting, and afterwards instituted cock-fighting festivals. These festivals the lads were expressly directed to attend in order that they might learn courage; a course approved by such moralists as Socrates and Solon. The Romans followed the same example. Of more primitive peoples, some of the earliest Chinese records mention cock-fighting; in India there are notices dating back to at least 1000 B.C.; and the Persians had practised it for centuries before the Greeks learnt from their example. Cock-fighting had also been traced amongst the Phoenicians, and some Jewish authorities believe that the Assyrian war-god Nergal was symbolised by a fighting cock, but this seems doubtful.

Antiquity of Cock-fighting

It is impossible to determine whether or not cock-fighting was introduced into Britain by the Romans. If it did not previously exist there, it certainly would be; but as Cæsar tells us "that the Britons *kept fowls for pleasure and diversion",* though it was unlawful to eat them, the probability is that they found this particular diversion already there before them, perhaps introduced by Phoenician traders. The first authentic notices of cock-fighting as an English sport only dates back to the reign of Henry II, and is by William Fitz-Stephen; but it is to be observed that this notice records it as fully recognised and carried on at public schools, especially on Shrove Tuesday. At a later period we find the "cock-penny" payable by each scholar at Shrovetide, in order to provide cocks for the customary festival, a recognised custom if not a regular fee.

Historical Details

For seven centuries cocking was more or less a national sport with us; and although disapproved and prohibited by Edward III and our eighth Harry, who was so tender-hearted as not to allow the poor cocks to be fought for the amusement of his belòved subjects, the latter built a cock-pit at Whitehall, to himself, wherein to take his royal pastime. Many succeeding edicts were passed against it, including one from Cromwell, the Protector, a fac-simile of which, with his seal, I have now before me. It dates 'Fryday,

March 31, 1654', and is signed by Henry Scobell, Clerk of the Council. On the other hand, many of our kings have been partial to and encouraged cock-fighting, and it has been called a royal sport. King James was one of its royal patrons; and in the Travels of Cosmo III, Grand Duke of Tuscany, through England, in the reign of Charles II, 1669, it is said, 'Attended by Lord Philip Neville, Gascoigne, and Castiglione, his Highness went out in his carriage to the theatre appropriated to cock-fighting, a common amusement of the English, who, even in the public streets, take a delight in seeing such battles, and considerable bets are made on them. To render the cocks fit for fighting they select the best of the breed, cut off their crests and spurs, keep them in separate coops, and mix with their usual food, pepper, cloves, and other aromatics, and yolks of eggs, to heat and render them more vigorous in battle. When they want to bring them to the trial, they convey them in a bag, put on artificial spurs made of silver or steel, and let them out in the place appointed . . . . This amusement was not new to His Highness, for he had seen it on board ship on his voyage from Spain to England'. The above description is not far different from later custom; and to the monarch already named is ascribed the introduction of the Pile cock, so called from his different and distinct colours.

"To such a height was this sport carried in former years, that in old deeds tenants were bound to walk so many fighting cocks for the use of the lords; and in corporation accounts of expenses I have seen large sums charged for entertaining this or that dignitary with cock-fighting. In the Easter week of 1822, in one pit, 188 cocks, weighing together 7 cwt. 4 lbs 6 ozs were fought for sums amounting to upwards of £6,000. Still more recently over 1,000 cocks have fallen in a single season in one of our northern towns.

"Victory lay with no special colour. In Queen Anne's time a noted sportsman, named Frampton, had the best strain of cocks of the day. They were grey, with a brown, tawny wing, and the progeny of 'Old Sour-face' was long in high repute. Greys, Yellows, and Red Piles were also highly prized, and Bradbury's Duns and Whites fought their way into notoriety. In the eighteenth century the mealy Greys, with black legs, beaks, and eyes, of Hugo Meynell and Sir C. Sedley could scarcely be surpassed. Then followed Mr Nunis's wonderful yellow Birchens, the Earl of Mexborough's true-feathered Duck-wings, Sir Francis Boynton's slashing Duns, and Col. Mellish's Dark Reds. Lowther's and Holford's Light Reds with yellow legs cut down everything before them; and Mr Elwes bred one of his red Duns that won twenty-seven battles. Then Vauxhall Clarke came into the royal pit to carry off the annual gold cup with his Greys. He bred different colours, and beating him was out of the question. The Cholmondleys, Raylances, Molyneuxes, etc., bred Smocks and the light Cheshire Piles, that would frequently electrify the pit by dropping cocks as dead as a log in a severe battle with the long odds against them. Dr Wing, of Leicestershire, bred all colours, and won with them. Sant's famous Derbyshire Dark Reds, with their dark-striped hackles, would always set the Derbyshire squires offering 100 to 80 on the battle; and old Nathaniel Monk, when sleeping in church at Dean, on being awakened by the beadle, cried lustily, 'I'll have the Black cock for a fiver!' so enamoured was he of the famous Black cocks of Lord de Vere. Mr Sketchley, the author of *The Cocker,* astonished the readers of sporting periodicals by the prowess of his Shropshire Reds; and Weightman, with his famous Parkhouse Reds, lowered the colours of the Lancashire men at Burton for the heaviest stake ever fought for; although it had been stated in error that Gilliver, when he won the main at Lincoln for £1,000 each battle and £5,000 the main , fought for the largest amount. The Earl of Derby, too, bred some grand black-breasted, white-legged Reds and Duckwings; and his Pile was looked on by admiring thousands, as the engraving was long exhibited in sporting print-sellers' windows. Dr Bellyse sometimes walked a thousand cock chickens out in a season, and was generally quite invincible. Once, on a sporting nobleman offering him £50 for a setting hen, he then and there lifted her off the nest and put his foot on the eggs; and on his lordship remarking that he bought the eggs too, he replied, 'If you had, I should have charged you a thousand'. His were about the only cocks that could beat Walker's celebrated Piles. I have not named a tenth part of the famous strains and breeders, but have mentioned sufficient to show that it was blood or strain that won, not colour; for even the Gurney Pied cocks were for a time thought to be superior to all others.

"The standard of a fighting Game cock is keenness of aspect, richness of plumage, and cleanness of feet. He must have a good boxing beak, very big, and crooked or hawk-shaped; large, full, fiery eye, and tapered head, not too long; for if the head be too long and beak straight, he loses much holding-power when taking hold to strike; long, strong neck; flat, broad body, tapering to wedge-shape to the tail; strong, long wings, so that when clipped the quills are of a powerful description; muscular, round, short thigh; legs (as to colour, I endorse the opinion of the most celebrated cocker of the nineteenth century, that the best he had ever seen were white, carp, and yellow, in order named as merit) of good hard bone, and not at all gummy or fleshy like other fowls, standing with a good bend at the hocks, so as to have a full spring when rising, and in line with the body, not out or straddling; spur set on very low down; clean, thin feet and toes, with a long, open black claw; and to be light, corky-fleshed, looking large to his weight. The great thing is breeding for *heel,* since it is the heel that always wins, and although health and strength is a great desideratum, without heel it is nothing.

"The old breeders never on any account bred from a cock or hen that was not in the most perfect health. The cock's feathers must not be dry or loose: he should be ripe in the feel, his flesh firm, and his crow clear. A general want of constitution requires no cross, the only cure is total eradication. Some were ever crossing with this fine cock or that grand hen, but the produce seldom came up to their expectations; and no one can dispute that the best strains of cocks ever bred, were bred in-and-in, and as soon as crossed with others, though equally good, were robbed of their winning qualities. Meynell's cocks as well as his hounds were so bred, and perhaps the world never saw either more perfect. Sant's were so bred. The celebrated Coath's had not a cross for forty years, and yet they were seldom beaten, and not in the least degenerated. Those Cheshire Piles and Bellyse's Reds that so often put down their opponents by a single fly, and were as much prized by some old Cheshire families as their own birthrights, were of one family, and as they were so long thought to be the very best, nothing else was allowed to contaminate them. In fact, with judicious care and a due regard to health and age, not only the best cocks, but the best horses, cattle, and dogs that England has been so justly proud of, have been bred in-and-in.

"They should early be put to separate walks that are healthy, where there is not a great number of hens, where they will neither be kept short of food, nor so hand-fed as to render them heavy and inactive; for the more exercise a cock takes in his walk, leading forth his hens etc., in search of food, so much more agile and active will he be in his battle.

"At two years old the cock is at his best for fighting; previous to which he went into the feeder's care, who reduced his weight, and got him into that high state of health and condition which centuries of close observation had brought to perfection. From very early times the craft have kept their secrets a profound mystery. At the middle of the eighteenth century, the cocking instructions of that scarce work, *The British Legacy,* published as a very special addendum, 'The following choice and valuable secret for feeding a cock for four days before fighting, which was communicated by a noble lord to J. Macdonald, M.D., by which remarkable and valuable method ninety-three battles have been won out of one hundred; now first published, by permission,' etc., etc. It was subsequently published in *The School of Arts,* and consisted of bread made of the flour of millet, rice, barley, vetches, cochineal, the whites and some yolks of eggs, wetted with ale, and baked for four hours, with which the cocks were fed after being purged, and with a slight allowance of bruised seeds and corn, and cooked flesh. Many other recipes for 'cock-bread' have been published at different times.

"Cocks put up and in training were weighed and their match colours and marks taken and noted three days before fighting, when each feeder would produce his birds as light as he could, and as soon as weighed and matched, proceed to get them up as quickly as possible to their highest weight. The birds would be occasionally purged if they were thought to require it, and at intervals muffles were put on and they were sparred a little for exercise and to keep them in good wind.

"Previous to fighting the wings were cut from the first rising feather slopewise. Hackle and cloak-feathers were shortened, the sickles all cut off, and the feathers around the tail, vent, and under the belly were cut short. The natural spurs being previously sawn off to about half an inch long, the silver or steel spurs were then placed on, and this again is supposed to be a great art by some; but I affirm it must be bred in the cock, and it is impossible to put on spurs to a bad-shaped cock to kill quickly, while a child can put on the spurs to a proper-shaped one. They must be padded firm in the socket, not tight, and rest well down on the leg, then tied tightly enough to prevent their moving, but not to cramp him, and from the natural spur place it in line with the outside of the hock. Then the handler stepped on the scene, who required a calm temper, as well as a quick eye and light hand, and had to take into consideration the condition of his opponent's cock as well as his own, otherwise he would not know when to force the fighting or when his bird required rest. This was the most difficult part of the whole routine of cocking; and Fisher, Straddling, Martin, Gomm, Probyn, Porter, and Fleming often won mains of importance by their exertions alone. Fleming was, perhaps, the cleverest setter that ever entered a cockpit.

"The Cockpit Royal, Westminster, was formerly the chief place for this sport, although there were many other public pits in the metropolis, and more than one of the London theatres was originally used for this purpose — Drury Lane for one; the next in importance was the royal pit, at Newmarket, immortalised by Hogarth. Hogarth is not the only one who has painted such scenes, as Vandyke, Elmer, Marshall, Barringer, Fielding, Alken, Cruikshank, and Wilson also painted them — the latter, a very large painting of the Salford pit and noblemen who frequented it. Neither were London and Newmarket the only places that supported cockpits, for few towns of any size were without one, and many cities and towns had established cockpits under patronage of their respective corporations; as an example, the Canterbury Corporation pit was an apartment of the beautiful gateway forming part of St. Augustine's Monastery, and this is not by any means a singular instance of the church and cockpit forming close alliance, either at home or abroad. A former venerable Dean of York bred such cocks as to trouble even Weightman with his best to beat them, and had he lived might have seen the old keeper of the York pit in charge of a Nonconformist chapel. On the site of the old Aintree pit there has been built a new church. The law, too, as well as the church, has been mixed up with this now tabooed sport; old gentlemen are still living who recollect, in attending to their professional duties at county sessions, having a six days' main of cocks fixed for the same time; and although the sessions might have been got through in the first two or three days, those magnates of the law would have been troubled with serious thoughts of having shirked their duties had they left before seeing the last battle in the six days' main decided.

"The Melton Mowbray pit, I believe, was the last built in England, at a cost of 700 guineas. The Subscription pit, at Chester, was one of the last abandoned, and no pit in England, perhaps, could boast of more aristocratic patronage, heavier betting, or superior fighting. The first main ever fought there was a main between Ireland and England of forty-three mains and ten byes. As the celebrated Doctor Bellyse represented Old England, he won, as he almost invariably did. The writer of this was, by the courtesy of the then occupier, forty years ago invited to see the spot where for years Ralph Benson's Shropshire Reds contended with the Piles and Reds of Cheshire to the admiration of all the old county families of Lancashire, Cheshire, Shropshire, and a great part of Wales.

Revival of Old English Game

This noble breed is now widely exhibited and still more widely bred, for its beauty and its quality of flesh. In combination of grace with agile strength it is unequalled, and the qualities for which it was bred produced also the utmost proportion of muscle (flesh) in the best places for the table, so that in wings and breast-meat it had no superior. It found its way into the exhibition-pen in the earliest days of poultry shows; but there, unfortunately, the changes of fashion of which other instances have been quoted already, played havoc with the breed. At first changes were slight, the birds being only slightly more tall and "reachy", which was generally admired; but the change went on, as shown in our next chapter, until the breed had been transformed out of all recognition. At last a reaction set in, and in 1882 a class for the "old" breed was offered at Cleator Moor in Cumberland, followed by classes with a special judge at Wigton in 1883; ever since which time, classes and popularity have increased continuously. In 1887 the Old English Game Fowl Club was formed, to encourage and watch over this noble breed, the first secretary to Mr J.W. Simpson, at that time of Silloth, in Cumberland, a county which was long the headquarters of the Old English Game.

Points of Old English Game

The points required in the Old English Game cock are a small tapered head, with a strong hooked beak, rather short and pointed, a quick, large, and fiery eye, the skin of face and throat of fine quality, loose and flexible; a rather long and very strong neck, a short flat back, wide across the shoulders, and tapering to the tail, which should be large, strong, and spread; breast large and wide, the pectoral muscles largely developed, and the breast-bone straight; the belly small and tight; the wings large, long, and strong; with short, round, and muscular thighs, and clean-boned, strong legs parallel with the body and well bent at the hocks. The spurs should be set on low, and be thin and sharp, the toes long, thin, straight, and spreading; the hind toe flat on the ground and extending straight backwards, the nails long and strong. His feathers should be hard, close, sound, and glossy, and his carriage should be proud, quick, graceful, bold, and smart. Such a bird should when taken in the hand feel firm in flesh and at the same time corky, or springy, and warm, drawing his legs well underneath him. He must be well balanced, and 'clever', that is, every part in due proportion, so that he sits easily in the hand, not lumpy or helpless; the thumbs should not sink in between the wings and the back, but be firm across; there should be very little fluff or underdown on him. His weight should be about 5½ lbs or 6 lbs, larger birds generally being coarse and dull, and lacking that alertness and quickness so desirable in this breed.

The hen should, as far as her sex will allow, possess all these points. Her head should be small and tapering, comb small, straight, and evenly serrated; very thick, curved, and pointed beak, large, bright, and full eye; strong shanks, the bone of fine texture, hard and evenly scaled, and if spurred, and of the same colour as the eye, beak, and plumage, it is a sign of purity of race and high breeding, flat clean feet, with long tapering toes, wings very long and full, with hard strong quills, tail large and fan-shaped, carried well up denoting strength and courage, as a weak low tail indicates weakness and a craven spirit. She should be wide in breast and back, and taper nicely to the tail; the importance of the hen cannot be overrated in producing Game cocks.

In colour of legs much latitude is allowed, the rule in breeding being that the eye, beak, and legs should match in colour. In Black-breasted Reds, for instance, white, yellow, carp, etc., are allowed to compete on an equal footing, each breed having its admirers, and being equally good and handsome. The white-legged ones may have also grey or daw eyes, as not only have some breeds (the Lord Derby's always had them), but they match the legs and beak, and the white under-plumage of these birds, which also usually show a few feathers wholly or partly white, in wings or tail, all in character with it.

The colours in Black-breasted Light Reds are much the same as in the modern Game, except that in the Old English the colours are richer and more brilliant, and may be darker, as some little latitude is allowed. For the hens, partridge colour is to be preferred to wheatens, the latter being used to produce bright hackle and saddle in the cocks, and if persisted in, also producing mealy breasts, while those bred from good partridge coloured hens produce sound coloured cocks, though a trifle darker in colour.

Brown-breasted Reds may have clear brown or robin breasts, or brown shaded or marked with black; the hackle and saddle are dark red or dark orange, and the eyes and legs dark. The hens to match them should be a rich dark mossy brown all over, or they may be black with a tinsel hackle; of course legs, eyes, and beak to match the cock.

The Silver Duckwings are in colour the same as the modern Silver Duckwing Game, or the Silver-grey Dorking, and should be kept to their own colour in breeding, and not crossed. The Yellow Duckwing may also be bred from Duckwings entirely, or can be produced by crossing with Black-breasted Reds, which will produce bright and rich coloured cocks, using either a Duckwing cock with a Partridge hen or a Black-breasted Red cock and Duckwing grey hen, though nearly all the pullets bred this way will show ruddy wings, which are fatal in the show-pen.

Duns (or, as they are improperly called, 'blues') may also be crossed with Black or Brown-breasted Red cocks; the former with a blue hen producing Dun-breasted Reds or 'Blue Reds', as fanciers are apt to call them, and a Blue hen with gold hackle and a Robin-breasted Red cock producing excellent 'Red Duns'.

Piles are much like the modern in colour, but brighter. Some prefer the white-breasted birds, but the streaky-breasted are also very handsome. The colour is liable to become lighter unless an occasional cross of the Black-breasted Red is used.

Blacks and Whites should, of course, be bred from pure self-coloured birds.

Spangles are very popular in the show-pen at present. They may be bred from Spangles, or as a cross with Black-breasted Reds, which also produce very good Red Spangles.

The Black-breasted Black Red is a breed that is considered one of the best and purest strains, and one that should be revived if not too late. The cock is a clear vivid dark red free from spot or streak, on hackles, shoulders, back and saddle feathers; while his breast, belly, tail, primary and secondary wing feathers, his thighs, legs, beak, and eyes are jet black, and his under-plumage black all over, also beneath his hackle; the hen to match him is a dark partridge, brick breasted, with hackles red above, and black beneath, and beak, eyes, and legs black also*. No breed was more celebrated than this in the old days, and it is much to be regretted if it is allowed to disappear from amongst us.

The Henny is also another most useful breed, being perhaps the best layer among Game fowls, and certainly in the front rank for the table. It is a very ancient breed of Game, and perhaps has been kept more free from crossing with other Game fowls than any other breed. The cocks are feathered like the

***Editorial Note: This description of the hen is now known to be inaccurate. A purple-black hen with dark-red striped hackle is the correct form.**

hens, hence its name, and the more hen-like their plumage, that is, the more rounded and free from sheen are the hackle, saddle, and tail feathers, the more are they entitled to claim purity of race. They are chiefly bred in Cornwall and Devonshire, though some celebrated birds have come from Wales. In colour there are dark partridge, red grouse, wheatens, greys, duns, blacks, whites, and spangles. The partridge and grouse coloured have generally pinkish white legs; these and the blacks are perhaps the most useful of all the breeds of Game. To some the lack of gaudy and shining plumage on the cock will make them appear plain, but there is a certain 'gamey' appearance about a good hen-cock that renders him very pleasing to the eye of one who understands Game fowls, and appreciates them.

There are very many other colours of Game fowls, but until they are more generally exhibited, a description of them is scarcely necessary. There are also strains of birds having peculiar marks, such as the 'Muffs', having a bunch of feathers growing beneath the throat, and the 'Tassels', having a tuft of feathers behind the comb, which may be either a few straight feathers in the cock and a small tuft in the hen, or something very much larger, as sometimes seen, and amounting to a large topknot. These peculiarities are pleasing to some people, and where they are followed by good qualities, it is well to preserve them with every care.

The Old English Game Fowl Club's Standard of points and colours seems to render further description unnecessary; but it may be well to give the proper way to describe the colour of a Game fowl. Experts always started with the breast, as in cock-fighting times this part was untrimmed and left intact; thus a black cock was described as a Raven-breasted Black, black eyes, beak, and legs; and a white cock, Smock-breasted Smock, white legs, beak, and eyes. Nowadays, a *Black-breasted Light Red* is often spoken of as a *Black Red,* a term at once misleading and ridiculous, and might mean anything, the black-breasted black red with crow-wings being the only colour to which the appellation black red applies; light red would be more intelligible. Usually the colour of some wild bird was used in describing the breast colour, as Raven-breasted, Throstle-breasted, Robin-breasted, or Pheasant-breasted, etc.

In rearing Old English Game chickens, it is well to keep them as dry under foot as possible at first. After a week or two they never grow better than if allowed to wander with their mother on a good range, and they will roost in the trees the winter through; but as it is not always convenient to allow them to do this, it is at least desirable to separate the cock chickens from their sisters at about three months old, when if put out of sight and hearing of any hen or pullet and under the charge of an old cock, they will usually run together peaceably, until they are ready to be separated to go on to walks. At five months old it is necessary to dub them. Snip off the comb and each wattle, taking care not to cut across the throat, and then take off the ear-lobes. The operation will not take half a minute, and the birds will eat directly after, showing how little pain is caused by this slight cutting, which saves the bird many a painful time, and often his life, when he meets a strange bird on his walk or gets out of his yard. The dubbing of this bird requires merely the removal of the comb, wattles, and ear-lobes for the bird's good; no severe cutting, trimming, or skinning is resorted to for appearance' sake, as in some modern breeds trimmed merely for appearance and to win prizes by unprincipled 'fakers'.

Plenty of pure water is absolutely necessary for Game fowls, and the evening meal should always consist of good, sound English grain: wheat, barley, and occasionally oats, and a few peas; maize is not at all a suitable food for Game fowls, being too fattening, and often producing 'scaly leg' and stopping the production of eggs.

This breed requires but little preparation for exhibition, as they never look better than when straight from a good range. They do, however, require to be tame, and should be placed in an exhibition pen a few times, and given some choice morsels to render them used to being handled, etc., as no judge can see the points of a bird if wild and crouching in his pen, or flying wildly against the top of it. If in good condition, just washing the feet, legs, and head will be sufficient, or should any feathers stick up each side of the cock's comb, they may be removed.

Judging Old English Game

In judging Old English Game, the first thing to be looked for is purity of race, gameness of aspect, cleanness and soundness of legs and feet, not to be thick toed, or with round fleshy shanks as are often seen, and his large fierce eye, whether it be dark, red, or grey. Then look to his shape, and then to his feather, if it is sound and glossy, elastic, and full of bloom, not soft, thick, or fluffy, as is often seen in inferior birds. Then take him out and handle him, and if firm in flesh and feather, clever in hand, with strong neck, and root to his tail, long and strong wings with sound unbroken feathers, and straight in

Figure 2.9 Game Cock Trimmed and Heeled

legs, breast, and back, and of good carriage and muscular, he will be fit to go into the prize list. Shape is the great point, for a badly shaped bird should stand no chance, and this cannot be told except by handling the bird, so that *no judge* will have done his duty until he has taken all the likely birds in his hands. Even then a person of experience is required to discriminate, for we often see birds with alien blood winning prizes at shows under incompetent judges. Feathery, fluffy Dorking types, and crosses with the modern Game, cannot be too carefully shunned by the judges. The great fear is that this grand breed may be again spoiled by exhibitors in a few years, by breeding to purely fancy points, until it becomes another edition of the modern Game, though bred on different lines. Let us remember that while the exhibitors were evolving from the same breed the modern Game, the old cockers, by judicious in-breeding, retained not only the Old English Game, but all its fine table qualities, its laying properties, and hardihood. They bred for purity of blood, shape, activity, hardihood, strength, and gameness, and it will be a standing disgrace if exhibitors allow themselves to lose all these useful points in breeding for the exhibition pen, and in seeking to improve upon a breed that was in its perfection nearly a century ago, and bred to a standard we can scarcely hope to attain nowadays.

HAMBURGHS AND REDCAPS

Under the general name of Hamburghs are now grouped a class of fowls which formerly were known under different names, but which share the common characteristics of rather small size, rose-combs, more or less white and round ear-lobes, slender, dark, clean legs, absence of the incubating instinct, and full sweeping tails. Looking at all that is really known of their origin, that of the spotted or Spangled, and of the barred or Pencilled varieties, appears quite distinct; the Pencilled Hamburghs having undoubtedly reached England from Holland under the name of Dutch Everyday Layers, and being also known as Chittiprats and Creels; while the Spangled Hamburghs emerged from Yorkshire and Lancashire, where they had been bred for years, from what stock no man knows, under the names of Mooneys and Pheasant fowls; the former from the round moon-like spangles, and the latter from the resemblance to pheasant-marking of their more crescentic spangles.

Figure 2.10 Hamburghs and Redcaps

Silver Spangled Hamburghs

The comb of an exhibition Silver-spangled cock should be even, firmly set on the head, long, and moderately broad, full of "work" or points, free from hollow in the centre, and ending in a long spike slightly pointing upwards. The beak should be horn colour, ear-lobes a clear white, smooth, and as nearly round as possible; face red, quite free from white; and eyes a dark hazel or red; legs a slaty blue. The neck should be nicely arched, with very full hackle falling well on to the shoulders; the breast full, broad, and prominent; back a moderate length, broad and level across, not round or up at one side; tail full, the sickles long and gracefully carried. Formerly the tail was liked rather high, though not squirrel-fashion, but is now preferred carried well back, as are all Hamburghs. The whole carriage should be graceful, jaunty, and cheerful. In regard to the plumage, the ground-colour should be pure silvery-white, quite free from straw or yellow, though at the end of the season, if exposed to the weather and sun, many good birds will turn rather yellowish. The head and hackle were formerly white; but the hackles and saddle feathers are now bred with black tips, the centres of which are dagger-shaped, while the fringe of the hackle extends the spot almost into a diamond. The saddle-hackles are the same, more heavily tipped, and the back is more heavily spotted still. Sometimes birds otherwise good will have hackles and shoulders with a little rusty or yellow: this is a great fault, of a worse kind than the slight tendency to straw which may be apparent in a good bird that has run long at large in the sun. The breast and thighs should be evenly spangled with round spangles, of a rich, satiny green-black, as large as possible so that they show the white; white throat or black thighs are the most usual faults here. The spangles which form the wing-bars must be especially even and distinct, as also those on the ends of the secondaries, which form the so-called "steppings" of the wings. Shoulders and wing-bows have dagger-shaped spots somewhat like the saddle-feathers, but shorter and broader. Each feather of the true tail and also of the sickles and side feathers, should be white with a large spangle at the end, those of the true tail being more of a half-moon shape.

Ideally the hen should correspond with the cock, with head-points in proportion, and white ear-lobes; but this is still an ideal, the combination of good lobes and perfect spangling being still an achievement for the breeders of the future. The spangles should be of an extrememly rich and satiny green-black, looking almost as if raised or embossed, and the back should be rather broad, so as to give room for them to show well. The breast must be spangled from the throat right round to the fluff, good distinct bars on the wings, and the tail clear, with a spangle at end of each feather.

Golden Spangled Hamburghs

The Gold-spangled Hamburgh is broadly similar to the Silver, substituting a rich golden-bay ground colour for the white; but there are differences in the neck and in the tail. The tail of the hen, and both the true tail and the sickles and coverts of the cock, instead of being a clear ground-colour with spangles at the tips, are of a rich and glossy solid green-black. And the hackles of both sexes, and saddle and back of the cock, instead of being spangled or tipped, or as near this as possible, are golden bay *striped* with glossy green-black; even the hen's tail-coverts being more a large black centre laced round with bay, than really spangled. The cock's wing-bow also, while now desired and standardised as deep bay with dagger-shaped tips to the feathers, as in the Silver, is still very commonly almost a self-colour, with little spotting at all; but this point is improving. The head points and body marking are as in Silvers, but the birds as a rule are much smaller. It is remarkable that the striped hackles and black tails in this colour, represent a marking almost similar to the old and now extinct *full-tailed* Silver Mooney breed, and was evidently one of the very oldest types of Spangled Hamburghs, and out of which the hen-tailed Silver Mooney was perfected.

REDCAPS

Most of the early poultry books describe amongst the varieties of Hamburghs a breed known as the Redcap, which is undoubtedly closely allied to one of those described under the preceding head. These Redcaps have also been known as "Manchesters", Moss Pheasants, and some other synonyms besides those mentioned below. Many years ago, as Redcaps, they used to have classes at the Sheffield shows, but subsequently dropped out, until more recently revived. They always had a very high reputation amongst those who knew them, as useful fowls, and the late Mr Hewitt wrote highly of them as such in

Figure 2.11 Silver Spangled Hamburghs

the first edition of *The Illustrated Book of Poultry*. He mentioned in particular the much better carcase, cocks reaching 7½ lbs to 9 lbs, while the meat was tender and delicate; and the fact that if equal weights of eggs from Spanish and Redcaps were used in custard making, the Redcap eggs went nearly a third further than the others.

The following short account of Redcaps is kindly supplied by Mr Albert E. Wragg, of Edensor, Bakewell, who has been largely instrumental in promoting the interests of the variety, and bred them for many years:

"The Redcap is one of our oldest breeds of fowls. It has been kept for a great many years in the counties of Derbyshire and Yorkshire, where it has always been most highly thought of, especially for its laying qualities. Recently, however, it has become better known, and it is now largely kept in the United States, Australia, New Zealand, France, Germany, and Belgium. It is generally supposed to have originated in Yorkshire, and is probably a near relation to the Golden-spangled Hamburgh, having been bred more for size and the large comb. Game blood undoubtedly enters into its composition, as the breed is a very pugnacious one, and a dubbed Redcap cock might almost be taken for an Old English Game cock. It has been known by many different names, such as Pheasant Fowls, Crammers, Copper Fowls, Yorkshire and Derbyshire Redcaps.

"The Redcap cock is a fine-bodied bird of noble appearance. Nothing could be more ornamental than his large, symmetrically-shaped comb, full of a great number of fine long spikes, with leader behind. It should be well carried, firm and straight, and standing well off the eyes. For years Redcaps were bred with very ugly combs — and some are yet to be seen winning prizes — and to this fact may be attributed much of the unpopularity of the breed in many parts of this country. The comb should be as large as can be comfortably carried by the bird. In size it should not greatly exceed five and a half inches in length and three and three-quarter inches in breadth. Birds with ugly combs should not be used for breeding. Frost seldom injures the large combs of Redcaps, for I have had birds roosting in plantations with 26^0 of frost, and not one has suffered in the slightest degree. The hen is a shapely bird, very active, and a good forager, and as a layer is second to none. She will generally lay from 150 to 200 eggs per year. Redcaps are long-

Figure 2.12 Derbyshire Redcaps

lived birds, and very hardy, and hens three and four years old will frequently lay as many eggs as a pullet. The eggs are white, or slightly tinted, very rich, and of a beautifully delicate flavour, and of good size, weighing about 2 ozs. The hens rarely go broody".

Silver Pencilled Hamburghs

Pencilled Hamburghs are smaller and lighter in make than the Spangled, and the Silvers and Golds were formerly known in Lancashire as Bolton Greys and Bolton Bays respectively, while the Silvers were, and still are sometimes, called Chittiprats. They are very obviously closely allied to the somewhat larger but singled-combed breeds described in a subsequent chapter as Campines and Braekels; indeed, a rose-combed Campine would be almost exactly a Pencilled Hamburgh of fifty years ago, but long breeding has reduced the size of that bird, and refined the pencilling of the hens, and altered that of the cocks, till the appearance of both has been considerably changed. The cock depicted in the early poultry books is however evidently pencilled on the body, very much as the Continental breeds just mentioned are at this day.

The head-points of the Silver-pencilled Hamburgh cock resemble those of other varieties, except in greater neatness and delicacy of appearance, the comb being somewhat smaller. The head, hackle, back, saddle, breast, and under-parts should be a clear silvery white, free from straw-colour. The true tail feathers are black, the sickles and side feathers rich glossy green-black up the centre, only edged with a fine white lacing, as sharply defined as possible. The marbling and splashing or grizzling with white which once was common, is impossible now for a successful bird. The wing-coverts or bar-feathers are generally more or less coarsely pencilled on the upper or invisible web, the tips sometimes showing a slight line of black across the wing; this slight bar was once cultivated, and is still allowed, but a white wing is now preferred. The secondaries are also usually black or coarsely pencilled on the inside web, but this is invisible. Formerly these feathers were black on the outer web except for a narrow band next the quill, but this dark colour is now discouraged. The fluff on the thighs is also now preferred as white as possible, whereas some pencilling used to be bred for there also. All these changes have been in the direction of breeding a whiter cock than formerly.

The pullet's hackle also should be silvery white and clear, but near the bottom is hardly ever so now. The rest of the body should have each feather distinctly pencilled across with narrow bars of black, as distinct and clear as possible upon the white ground (or in the case of Golden Pencils, golden ground), and as *straight* across the feather as possible. The pencillings should show a rich green gloss, and range as much as possible into lines round the body, as in what are termed "ringlets" in Plymouth Rocks. The finer the pencilling, and therefore the more numerous the bars, the better; and the marking should extend from under the throat to the end of the tail. On the breast the pencillings will be fewer, and under the throat is a particularly weak place, very apt to come merely spotted, or with horse-shoe markings: but some birds are well marked even in this region, though not so well as elsewhere, and though the best ones are generally most marked on the hackle. The fact is that breeding for pencilling alone always tends to produce pencilled hackle also, as we have seen already in Partridge Cochins and Brahmas. The tail should be well pencilled straight across, and this is not so very rare in the two top feathers of the tail itself; but it is curious that some pullets properly pencilled there will fail in the longer tail-coverts, and *vice versa,* so that a fine all-round complete tail is rather rare.

Golden Pencilled Hamburgh

The Golden-Pencilled Hamburgh is in every respect save ground-colour similar to the Silver-pencilled breed. The ground-colour in the pullets should be about the colour of gold, as rich and bright as possible; the pencilling being exactly like that of the preceding variety, as distinct and yet as fine as can be got; that is, as many bars as possible across each feather, provided they are distinct, straight, and of a good rich black colour. The neck-hackle, like the Silver birds', should be clear. The cock is of a deeper tint, his colour being somewhat between that of his own hens and of a Red Game cock; it is described in the Standard as a red bay, while the pullets are termed golden bay; it must be neither too red nor too pale, but what might be called very rich in effect. His proper tail-feathers are black; the sickles and side hangers rich black edged with bronze or gold, the edging being usually rather wider than in the Silver-pencilled bird, but a fine narrow edging is much the best for a breeding bird, though not so showy in the pen. Sickles all black, or bronzed all over, with scarcely any black in them, are now out of the question, though at one time they were tolerated. Besides the quality of the black pencilling, one of the points in this variety is the evenness and richness of the ground-colour. Some pullets, otherwise good, are uneven in this point, the ends of the feathers being a lighter colour than the other parts. As summer advances most birds fade in colour from the effects of the sun; but some pullets of a good rich ground retain this much better than others, which is, of course, a great point in their favour. In the cocks the same fault is commonly seen, appearing in the shape of a lighter shade on the tips of the feathers on the breast and upper-parts. This fault is to be particularly avoided.

Black Hamburghs

Under the name of Black Pheasants, the Black Hamburgh was known and exhibited a hundred years ago at the Lancashire village shows already alluded to. In those days it was of unmixed Hamburgh blood, and the opinion of the late Mr Teebay and some other old breeders was, that it was originally bred from the full-tailed Silver Mooney, many chickens of the latter coming almost black. The birds of that day were shorter in the leg than now, and of the same shorter formation generally as the older Mooneys, and with ear-lobes much smaller and coarser than now. This old breed was undoubtedly crossed with the Spanish, in order to improve size and quality of the ear-lobes; and the size of the fowl and of its egg were both improved at the same time, while it likewise became more suitable in disposition for small runs. Unfortunately, the cross also introduced a considerable tendency to white in face, and some coarseness about the head, which have required a great deal of breeding out; the form of the bird also became somewhat taller, though anything like the stilty carriage of the Spanish is most objectionable, and the usual Hamburgh carriage should be sought as far as possible.

The Black Hamburgh of the present day is a most striking bird, the combination of bright carmine head and face and comb, smooth white kid ear-lobe, and lustrous green plumage, making a most beautiful whole. The ear-lobe is considerably larger now than even twenty years ago, that of the cock being about the size of a florin; it should be round in shape and smooth all over, perfectly free from folds or lines or creases. Such a lobe is, of course, very apt to be accompanied by more or less white in face, especially under the eye, a fault much more common in Blacks than in any other varieties of Hamburgh. Very few old birds are indeed free from it, but in young ones it is a serious fault. A gipsy tint sometimes seen is also disliked, a cherry red being the desired colour. In combs also there has been a perceptible change in fashion, in the direction of a longer spike or "leader" at the back, especially in the cock. Both sexes are of rather larger size than other Hamburghs, except perhaps some of the larger Silver Mooneys, and the cock is of somewhat more commanding carriage, and should be particularly long and full in feather about the saddle and tail, the sickles and side-feathers being broad, but as sound and close in web as those of a Game cock. The shanks should be dark leaden blue, and are very often nearly black up to a year old, but after that usually get lighter with age.

HOUDANS

The best known in this country of the French breeds is the Houdan, which was briefly mentioned by Messrs. Wingfield and Johnson in *The Poultry Book* of 1853, as the Normandy Fowl, and fully described by Mr Geyelin in 1865. The economic value of this breed is very great, as it is very hardy, a quick grower in chickenhood, when well bred a first-class layer of good-sized eggs, and of very delicate flesh, with a tendency to fatten well. The male birds are also unusually vigorous and fertile, many of them requiring more hens than would be usual with other birds of the same size.

The Houdan cock should be as large as possible, the adult being 8 lbs or 9 lbs or more. He has a good-sized crest rather inclining backward, and a peculiar comb, resembling the open leaves of a book with a sort of mulberry in the centre, or a butterfly with open wings. This is large in the cock, but rather small in the hens. The wattles are moderate and rounded, beard and whiskers rather full. The neck-hackle is full and thick, the body very long and deep in keel, carried in a very sturdy manner, the tail full in sickles. The legs are clean, pinky white mottled with black in colour, and five-toed; the plumage black and white mixed, of no exact pattern, but about equally broken in colour when adult, and rather more black than white when young; usually more or less approaching a crescentic white marking on beetle-green centre,

Figure 2.13 Black Hamburghs

over the breast and body. The hen has a fuller, globular, and very Polish crest, with much smaller comb, her weight 6 lbs or 7 lbs.

Houdans develop quickly, and cockerels at six to seven months are usually in full feather; pullets at from five to six months, at which time they generally commence to lay. They fatten very quickly, and if over-fed about this age it retards their laying. They stand confinement capitally; indeed, exhibition birds do better if confined during the show season than if allowed to run out.

In breeding Houdans, I may at once remark that both sexes of the highest excellence can be bred from the one pen. No need, in Houdans at any rate, of one pen for cockerel-breeding and another for pullet-breeding. In breeding for exhibition birds, nice medium-coloured birds should be selected, the black a good solid green-black, nicely broken. They should have full crests, as smooth — especially in front — as possible, and neat, even combs of butterfly pattern. The comb may vary in size and shape, just as the wings of different sorts of butterflies do, but the shape and pattern of the comb must be the butterfly, with the wings open, or nearly so. They should have light or whitish legs and feet, mottled with blue or blackish blue. If the white is a pinkish white, all the better. If a pen of birds of this description, with good deep square bodies and of a good strain, be bred from, they are pretty sure to produce a good proportion of chickens fit to show. Some chickens come with very dark or nearly black legs. These usually are very dark in plumage, but not invariably so. These black or dark legs always change with age to blue or bluish mottled colour, and though the black or very dark leg is much against a bird in the show-pen, the blue or blueish mottled colour is very little detriment indeed to it, and such birds frequently win the highest honours. Those who happen to have dark birds or light birds of much excellence in Houdan character and points, and who wish to breed from them, must select light hens for a dark cock, or, which I much prefer, a light cock for dark hens. Mated in this way, a fair proportion of good coloured chickens will be the result. Any foot deformity in the stock birds is very likely to be perpetuated, and whether these be dark, light, or medium-colour, let me repeat that the black in the plumage must be a good sound green black.

Two-year-old birds are best to breed from, as the produce have greater robustness and usually attain greater size; but year old cocks mated with two and three year old hens give excellent results. Singular to say, the largest hen I ever bred was from a pen of pullets mated with a two-year-old cock; but I should not expect this result to be repeated. Houdans are long-lived. I have several times shown a cock up to five years, and hens up to six and seven years, and on one occasion I bred from a cock five years old, and he bred freely and well in March, and this, too, after a fairly long show career.

MODERN LANGSHANS

The following notes are kindly supplied by Mr F. Onslow Piercy, of The Elms, Lowthorpe, formerly one of the most successful exhibitors of the modern type of Langshan, and whose remarks upon its economic qualities are especially worthy of attention.

"Perhaps a few remarks on the Society type of Langshan from one who has bred and studied the breed for many years, may possibly be of some use to persons who contemplate going in for that type. In the first place I ought to say that originally I produced my Society type of Langshans from the original, by careful selection in picking out the tallest, finest boned, closest feathered, best coloured, and most stylish looking birds, and breeding from them. Most breeders will be aware that many different types may be produced from any breed; it is only a matter of time and selection in mating the breeding pens, choosing the birds most likely to produce the type you are aiming for. Briefly, the difference between the original type as imported from the Langshan district, and the modern type, is this: the former is shorter and coarser in bone, much looser in feather, carries much more fluff, and is coarser in feather on the shanks and toes; also the feather about the hocks is very much heavier; the body is shorter, and the colour too is not so brilliant as the later type of Langshan.

"The latter is a tall bird, with a nice length of shank but medium length of thigh, sufficiently strong in bone to carry the weight of the bird, neither too coarse or too fine; good length of body, with a long, deep, and well rounded breast. It is close in feather and of a brilliant green colour, free from purple; a tall, stately, deep-breasted bird, with a beautifully rounded outline and good upstanding carriage, in proportion all round, and not a bird that strikes one as being excessively leggy, neither should it have a cut-away appearance in the breast, as some specimens have. I consider there is nothing handsome in a

Figure 2.14 Langshans

***Editoral Note: This would now be regarded as the Modern Langshan.**

Langshan if you can almost draw a straight line from the head right along the keel without catching the breast. The scales on the shanks, etc., of young Langshans, till after the adult moult, should be nearly black, turning paler afterwards. The shanks and outer toes should be nicely, although not too heavily feathered, the shoulders should be broad, the shanks set well apart and the tail carried slightly elevated, and in the case of the cock should have an abundance of green side hangers, and of course the two sickles projecting beyond the rest; the eyes dark, face and comb brilliant red; the latter firm, erect, and evenly serrated. The cock should be fairly long in the neck, with a full and well rounded hackle. Nearly all of the best Langshans show red between the toes and the scales down sides of shanks, especially the male birds.

"In breeding Society type Langshans, I prefer to breed from two-year-old hens, although I prefer to have bred some of my best pullets from first season hens. I would choose large but not coarse boned hens (there are many modern Langshan hens weighing over 10 lbs each, and not a bit coarse in bone for their size), as green as possible in colour, with shanks well apart, broad across the shoulder, and with a long deep breast and good length of back, close in feather and of good carriage, well, although not too heavily feathered on the shanks and outer toes, and as dark as possible in the eyes; with neat, small, firm, erect, and evenly serrated combs if for pullet breeding, but for breeding cockerels it does not signify so much if the breeding hens' combs are large and loose, so long as they are evenly serrated.

CROAD LANGSHANS

In previous editions of *The Book of Poultry* this important section of the Langshan family has had to be content with but scant notice beyond the discussion in the opening pages of this chapter as to the causes which led to the production of the two types of the breed, now known as Langshan and Croad Langshan. With the present wide popularity of the latter variety, due largely to its utility properties and also to the better spirit now prevailing amongst all sections of Langshan breeders, we have thought it desirable to give Croad Langshans separate treatment, and Mr Herbert P. Mullens, Hon. Secretary of the Croad Langshan Club, has very kindly contributed the following notes on the breed:-

"The Langshan fowl was first imported from the Langshan District of North China by the late Major Croad, of Durrington, Sussex, in the year 1872, and at that time there was much discussion as to whether it was a distinct and pure breed, but eventually it was clearly proved that the new importation was not only a distinct breed but one of exceptional merit for utility purposes.

"There were two distinct types, one tall and reachy and the other of medium height with full flowing tail, and these two have been accentuated by their respective admirers.

"It is with the latter of these types that this article deals — namely, the Croad Langshan; the prefix 'Croad' was given to this branch for two reasons, firstly, to distinguish them from the tall, reachy type, and, secondly, as a perpetual reminder of the debt that poultry keepers owe to the late Major Croad for introducing such a useful and ornamental breed, and to the late Miss Croad for the way she championed the breed through good and ill report.

"In colour the Langshan is a dense black throughout, with a brilliant beetle green gloss upon the plumage; a flat and broad-shouldered bird, with a deep, long meaty breast supported by legs of medium or sufficient length to give a graceful carriage to the bird; the legs are feathered down the outer sides and toes, but not heavily. Adults weigh about 9 lb in cocks and 7 lb in hens. They are very good table birds, possessing flesh which is exceptionally fine in grain and very juicy, very much resembling that of the Houdan; the skin is very white and thin. But their chief call for public favour is their exceptional egg producing qualities, especially during the winter months; the egg is brown, varying in tint from very dark to a medium shade. Some of the eggs are far darker than those laid by any other breed.

"As an exhibition bird the Langshan first appeared in separate classes at the Crystal Palace in 1876; and though for some years the Croad Langshan was seldom seen at shows, of recent years, since the formation of the Croad Langshan Club, it has been steadily advancing in public favour, both as regards numbers and quality, until now at all the big shows well filled classes are to be seen. In a nutshell, the Croad Langshan is of such all-round excellence that it must always appeal both to the show lover and utility fancier."

Figure 2.15 The first or Original Langshans

LEGHORNS

POINTS AND QUALITIES OF LEGHORNS

The Leghorn family constitute a group of one of our very best and most profitable laying fowls, though unfortunately some changes that have taken place since their introduction have by no means tended to increase their value in this respect. They have the large single comb of the Mediterranean group, straight and upright in the cocks and falling over in the hens, almond-shaped white ear-lobes, with red faces, and the general type of the class. Their chief characteristic differences are their bright yellow legs, rather smaller size, sprightliness and activity, and greater hardiness. When first imported, the tails were carried very upright, or even squirrel-tailed, which had been the fashion in America; but as we predicted and advocated from the first, that fashion has never been approved in this country, and is now abandoned in America also. Another change is more regrettable. The original Leghorn comb, though of the family type, was moderate size, and thin and fine in quality: there has been too much tendency towards a large and beefy comb, which has been deplored by every practical writer upon the breed without exception, and has necessitated the wholesale dubbing of breeding stock. American breeders have fortunately never adopted this fashion, which has gravely affected the egg-average of some strains: and it is to be hoped that some little reaction lately observable may continue, and the more moderate American and Italian comb again prevail. Breeding for size has also, in some cases, been carried too far, the largest birds by no means laying the largest eggs, and being inferior in activity and hardiness to those of more typical size.

Leghorn chickens are very hardy, and feather easily. Some of the cockerels weigh 6 lbs and few are under 5 lbs, and the flesh is by no means bad eating, being more juicy than that of the Minorca. The great usefulness of this group is however as exceedingly hardy, non-sitting, laying fowls, whose eggs are large for the size of the bird, even small Leghorns rarely laying eggs less than 2 ozs in weight, and many decidedly heavier, though (as we just observed) the largest by no means lay the best ones. Both Mr Tegetmeier and ourselves reported upon the first specimens received as amongst the very best layers we had met with; and except in some strains which have been injured by crossing for show points, this character has been preserved, and is always easily bred up to. The White Leghorn is renowned in the United States and Australia as a breed which, bred for laying and adequately fed, is easily got up to an average of close upon 200 eggs per annum, while these eggs are of a good marketable size; it has the current name there of being "the business hen". It has the further valuable property of maturing early, and at a very uniform age, so that if adequately fed pullets may be depended upon to lay before six months old. They can be forced much earlier, but this is not advisable, and it is far better to date the time of hatching accordingly.

White Leghorns

Of all the varieties of Leghorns, the White has been longest known in England. It has become much larger than when first imported — many think too large — and it has suffered as much as any from large overgrown combs, which bow down the necks of the poor birds in too many pens; but we have said enough on this point. Mr F. Tootill, has kindly supplied the following notes on this variety:-

"It is with pleasure that I consent to supply a few notes respecting the leading and most popular variety of Leghorn. In doing so my desire is to assist the amateur and breeder, and to further the interests of the breed; and with this end in view any little knowledge I possess is most freely given, and I am in hopes that from the perusal of these notes those for which they are intended may derive benefit.

"During the past few years the White Leghorn has been much improved, to such an extent that at present shows it is no uncommon thing to find specimens staged which are equal, both in size and head-points, to our best Black Minorcas. In fact, this has been carried so far that it has been questioned, and has been the cause of long and heated controversy, whether breeders were not now losing that beautiful stately carriage of the typical Leghorn in their desire to obtain size. Size is difficult to secure, and, when once obtained, a breeder has reason to be pleased with the result of his exertions; but we must have *type* in conjunction with it. Experience tells us, however, that it is little use nowadays to exhibit the small 'pretty' whites, our judges signifying their requirements by their invariable decisions in favour of size, sometimes in preference to head-points and general quality. This applies more particularly to the all round judge, and is to be regretted, as type and quality should always have prominent consideration. Only on one occasion, in my experience, has any specialist judge been so infatuated with mere size as to have given the preference to a huge body and ungainly carriage against a perfect head, with good shape, on a smaller body; but as this was at one of the most important shows of the season, such a decision may have a considerable effect on breeders. The position of judge is as important as the task is an unthankful one, and specialist judges at our principal shows should not lose sight of the fact that they have, to a certain extent, the destiny of the breeds on which they adjudicate in their hands. In White Leghorns, some years ago, there was some foundation for the view that in consequence of the tendency above noted we were threatened with loss of type, the all-important feature. I am glad to say now, however, that breeders are paying much more attention to shape than formerly.

"Although it is a characteristic most difficult to put down on paper, the shape of a Leghorn is totally distinct from that of a Minorca, just as the Andalusian is. In breeding I make this feature as important as purity of colour. I place vital importance on type and colour, because a Leghorn ceases to be a Leghorn when it is not typical, just as it ceases to be a *White* Leghorn when the colour is impure. It has been suggested that, to secure the size of present White Leghorns, foreign blood has been introduced, such as that of the White Rock, White Malay, etc., etc.; but this theory loses weight when we look at the perfect head-points which have been shown simultaneously with increased size.

"The time is long past, too, for one to be able to win in anything like decent competition with Leghorns of the straw-coloured variety, pure colour being demanded by every judge. This can only be obtained by breeding, discarding *all* inferior-coloured specimens (no matter what other good points they possess) when selecting one's breeding stock. If any degree of success is desired, this, as I said before, is specially important, as when a Leghorn is not a pure white it is not a *White* Leghorn.

"The White Leghorn has seen improvement in more characteristics than size and head-points. We now have the cocks much more heavily clothed in feather than in former years. Few are exhibited today with scanty hackles, and close whip tails with narrow sickles. A White Leghorn requires to be furnished with long, flowing sickles and secondaries to be a thing of beauty. Scant feather looks particularly amiss on the larger specimens, and strange to say these are generally the ones that are deficient."

Brown Leghorns

The colour of the Brown Leghorn as first bred in America was very uncertain, some of the first imported specimens having a great deal of brown in the breasts of the cocks. This was however gradually bred out in favour of the black breast; and for many years the accepted colour has been what may generally be described as that of the black-breasted red Game, with the exception that the cock's hackles are somewhat darker, or orange-red, and should have a little black striping near the shoulders. This colour resembles that of the Game cocks before these were bred quite so bright, and there is no doubt that several crosses with black-breasted red Game were employed at different times, to improve and fix the colour and marking. The effects of this crossing are still seen occasionally in pullets with dark legs and feet (from the willow legs of the Game), and as this fault is especially obstinate, specimens which exhibit it should be carefully avoided in breeding. It is less common now than a few years ago.

Pile Leghorns

Pile Leghorns should probably come next in date, having been produced by crossing Whites with Browns, in the same way as were first produced Pile Game. It was in 1881 that Mr G. Payne mated up his first pen of White and Brown; but it was not until January 1886, that he was able to exhibit two pullets and a cockerel, the latter being poor, but the pullets good. At the Dairy Show of that year, however, he produced two pairs of Piles which left little to be desired, and took first and third prizes in a mixed class; and for a time the colour was fairly popular. Other crosses have been introduced by various breeders at one time or another. The article below speaks of a Game cross, and its effects on the type of the strain; and perhaps one of the most remarkable "flukes" in the history of poultry-breeding was the colour and success of some Pile Leghorns exhibited in "the nineties", which the breeder himself stated to be bred from a cross-bred bird deriving parentage from the Light Brahma! This cross probably accounts for the feathered legs seen on a cockerel at the Palace Show in 1894; but we have a vivid recollection of the colour shown by this exhibitor for one or two seasons, which was marvellous. There can be no real occasion for crossing any further.

Duckwing Leghorns

The colour of Duckwing Leghorns is in all but one point practically the same as in the corresponding varieties of Duckwing Game. That point is the striping of the hackle: as the Brown Leghorn is a striped breed, so the Duck-wing varieties have the longer feathers of the hackle somewhat striped also. Mr Payne had made no attempt to breed Golden and Silver strains, but as the variety was bred more generally this became inevitable. A good gold-coloured cockerel almost bred pullets red or rusty on the wings; hence pullets had to be bred from lighter or more silvery cocks. And conversely, good-coloured Gold cocks could only be produced from more or less rusty females. Both classes are now recognised by the Standard, and are necessary for breeding, but at the majority of shows, where there is one "Duckwing" class only, the winners are usually Golden-Duckwing cocks, with almost silvery hens.

Breeding Duckwing Leghorns

The colour of Duckwing Leghorns is in all but one point practically the same as in the corresponding varieties of Duckwing Game. That point is the striping of the hackle: as the Brown Leghorn is a striped breed, so the Duckwinged varieties have the longer feathers of the hackle somewhat striped also. Mr Payne had made no attempt to breed Golden and Silver strains, but as the variety was bred more generally this became inevitable. A good gold-coloured cockerel almost always bred pullets red or rusty on the wings; hence pullets had to be bred from lighter or more silvery cocks. And conversely, good-coloured Gold cocks could only be produced from more or less rusty females. Both classes are now recognised by the Standard, and are necessary for breeding, but at the majority of shows, where there is one "Duckwing" class only, the winners are usually Golden-Duckwing cocks, with almost silvery hens.

Silver Duckwings, so Mr Hinson wrote us, are usually bred from one pen, the same mating producing both sexes good if the strain is well bred, and the colour and markings sound on both sides. Where this is not so, somewhat inferior colour in either, or in both, often breeds very fair pullets, though failing in cockerels. Pure silvery white in the hackles of both sexes is the great criterion. The best mating of all is that of a silvery-hackled cock with a rather dark grey but absolutely pure-coloured hen.

To breed Golden Duckwings, two pens are practically requisite, though not so much so as before the rich golden wing-bows now sought in the cock, had replaced the deep maroon or crimson once fashionable. For cockerel breeding it is best to select a typical Golden exhibition bird, sound in all his colours, and put to him hens with rich salmon breasts, and wich may with no detriment have a little warmth or rust on the wing. For breeding pullets, the cock should be bred from Golden pullets, very sound in his black all over, but rather light on shoulder, and is none the worse if rather broken in colour there: if his hackles also tends to being silvery it is all the better. His mates should be pure in colour, as near as possible to ideal exhibition hens. If at ay time too much colour comes in the hackles of either sex, or the bodies of the hens, a cross of Silver Duckwing blood is desirable.

Buff Leghorns

The Buff Leghorn cock should have the same kind of comb and colour of beak and legs as the Brown or White varieties, whilst his plumage colour should be either a lemon or orange buff, the breast feathers being a little richer in tone than the back, but certainly not in any great degree such as to form a decided contrast. The whole plumage, whether of the lemon or orange shade, should be quite even and free from mealiness, and the tail should be solid in colour, perhaps a little deeper in tone, but free from white or black, or partly white or black feathers.

The Buff Leghorn should possess the same characteristics as to head points, style, shape and colour of beak, lobes, and legs as the Brown hen, whilst her plumage should be an even shade of buff all through, without variation in any part, though in many specimens the hackle will be found to be a shade or two deeper in tone than the body.

Figure 2.16 The Jungle Fowl and some of its Descendants

Undoubtedly the infusion of the foreign blood already mentioned will make itself apparent now and again, and the reappearance of slight feathering on the legs, red in lobes, and white in flights and tail, are to be expected; but by very careful mating up of the breeding pen and persistent weeding out of the faulty specimens these evils will be overcome, and the certain production of solid coloured Buff Leghorns will be established.

Breeding Buff Leghorns

In selecting Buff Leghorns for breeding, great attention should be given to the head points of both the male and female, for up to now these have been comparatively neglected, the chief aim having been to produce uniformity of colour. Now that this has been established, the improvement of comb and lobes, especially the latter, requires earnest consideration. The stock cock may be a little deeper in colour than the hens that he is to be mated with, but he should not be too dark, particularly if the hens are inclined to be a light buff, for the extremes of shade never amalgamate well, and the progeny are apt to become mottled when they assume their adult plumage. Whether the birds be lemon or orange buff, they should all be about the same shade, excepting as stated above, that the plumage of the cock may be a little richer in tint. Hens that show any amount of mealiness should be discarded, as should also those that have decided black ticks at the ends of the hackle feathers, or on the tail feathers. Too much stress cannot be laid on the under-colour, and every stock bird should be closely examined to see that the buff extends well into the under plumage; for very often though the surface may be all that is desired, the fluff will be found to be quite white. Should the buff extend almost to the skin, there is but little fear but that the progeny will come sound in colour throughout. The tail of the cock will probably prove the greatest difficulty, this being especially the home of white and black feathers. Though both are faults, yet the former is the greater, and likely to be reproduced in a larger degree than the latter. Still, as really sound coloured tails are even yet the exception, choice should be made of the bird that has the least white in the main feathers of his tail.

Black Leghorns

The Black Leghorn as a variety ought to be better known for its qualities than it is at the present time by the majority of poultry fanciers. It is an exceptionally hardy bird to rear, and bears confinement well, and is a splendid all-round layer. I have had pullets, hatched in April, commence to lay at four months old, laying a good saleable white-shelled egg, and continuing to do so through the winter months. Frost and snow seem to have no effect, the birds being out both day and night all through. I have had young cockerels to crow at 39 days old. They are very active, and the birds at liberty are splendid foragers and small eaters. Fifty pullets or hens in a field make a splendid sight to see, with their jet black bodies and bright yellow legs. They always seem to be on the alert, and will take wing at times for fifty or seventy yards. They are rather of a wild nature when at liberty, and scarcely approachable. I firmly believe that a hundred and fifty Black Leghorns will equal two hundred of any cross-bred birds brought to compete against them; that is, for egg production. I have also had cockerels weigh 5¼ lbs at five months old, and the flesh is white and juicy.

The chickens are easy to rear, and free from disease. They are generally dark, with white underparts when hatched, the majority then having dark legs, but becoming yellow as time goes on, more so with the cockerels than with the pullets. The chief difficulty in the cockerels is that they are subject to white in their tails- but this is greatly improving. I have possessed birds with sound black tails, but they have been wrong in other things, such as being inclined to show red feathers, or bronzy looking on the back. I always try to breed from a cock bird that has the least white in tail, but still having a good sound rich black body colour and good yellow legs. The bigger the lobe is the better, as we are still wanting in this particular point, especially in pullets, but the birds are as a rule quite sound in face.

In mating a pen of Black Leghorns together, I should advise to get a big sound-coloured bird, with good face, ear lobes, and leg, and as little white in tail as possible. A cock of this description mated to six pullets with good sound black bodies and good yellow legs, and good lobes and nice folding combs, will breed birds for prizes in the show pen.

As to the Standard for Black Leghorns, the plumage should be a rich glossy black, free from feathers of any other colour; the more sheen the better. Legs and feet yellow; eyes bright red; beak and toe-nails yellow or horn-colour; comb, face, and wattles a bright red. Size as large as possible.

Cuckoo Leghorns

Cuckoo-coloured or blue-barred Leghorns are occasionally seen, but are not popular. They appear to have come chiefly from the Continent, and to have occurred naturally from mixture of black and white, as so many other similarly-coloured races have done. There is no doubt that they are just as good in qualities as other Leghorns. But the colour does not seem in this variety to be attractive, and, there being no adequate support for it in classes, it is little cultivated. It is a difficult colour to produce in this breed, white and other foul feathers constantly occurring. This difficulty is not now found to the same extent in the barred Rock, which has been bred for colour and marking through many generations, and in large numbers; but in a variety so little bred as the Cuckoo Leghorn it is felt severely. With such stock as is obtainable, we should advise selecting two-year-old birds (that age often showing up faults not seen in chickens), weeding out severely for any foul feathers, and selecting the medium colour and barring in both sexes. Strains are not fixed enough to breed by the same rules as barred Rocks, and this course will be found on the whole most successful.

Mottled Leghorns are also rarely seen, but are likely to be displaced by the Ancona, if not practically the same bird.

American Leghorns

The Leghorn is bred and kept in America and also Australia to an extent quite unknown in England; but the birds there generally are different from the English style, far nearer the original type, and, as a rule, more prolific. In one respect English type has prevailed. All the first birds sent to England were very high in tail but we from the first advisedly opposed that style, as sure to be fatal to the fowl if adhered to, and the result has fully justified our action, which happily proved decisive at a time when the question hung in suspense. On both sides of the Atlantic the flowing tail is long since fully recognised. But the American differs in other respects. It remains considerably smaller than the Minorca, and is also rather higher on the leg than the English, of rather more slender form and sprightly carriage, and with the much more moderate comb of the original bird.

MALAYS, ASEEL AND INDIAN GAME

MALAYS

A bird similar to the Malay in all essential characteristics is still the indigenous or common fowl of India, as well as the smaller peninsula whose name it bears, and so far its origin is not difficult to determine; further back the problem is not so easy. The Malay characteristics are very distinct and peculiar, consisting more in general points than any fixed standard of colour.

Many years ago, when Cochins were unknown, Malays were the only Asiatic breeds which could be used to give size by crossing to smaller fowls; and being then freely imported, were large heavy birds. But of late the superiority of the Cochin in temper has so diminished Malay admirers, that there has been little demand for imported birds; in-and-in breeding has been the necessary consequence; and they now usually appear in the pens as actually **small** fowls, though their real weight is always greater than appears. The cocks used to stand thirty inches high, and weigh eleven to twelve pounds, the hens in proportion; but now at the least one-third must be taken off these figures.

Characteristics of Malays

There is perhaps hardly any breed with characteristics so distinctive and well marked, which makes it the more surprising that some of the "all-round" judges should appear still unable to grasp them. The head of the cock is large, and particularly very broad, with heavy overhanging eyebrows, which give a cruel expression to the face by no means belied by the character of the bird. Besides this, some of the older writers describe the bird as "serpent-headed", and observation will confirm the singular aptness of the expression, quite different from the snaky head sometimes spoken of in Game. The beak is very stout and quite curved, almost hooked in fact. The face is smooth and skinny, with the throat rather bare, wattles and ear-lobes small, and the comb unique, neither single, nor rose, nor triple, but like half of a walnut covered with very small projections. This should be fairly small and set well forward; but if two small combs are bred together, it is significant that *pea-combs* are apt to result, showing clearly the relationship with both breeds treated of in the following sections. The neck is long, and hackle full there, but short and scanty below, which gives the entire neck somewhat the appearance of a pillar the same diameter all the way down, and rising almost abruptly out of the shoulders. The body is large round at the shoulders, which are very prominent and carried high, and tapers away towards the tail, the back being long, rather convex in outline, and slanting downwards: the tail is also carried low and drooping, so that the back of the neck, the back and top feathers of the tail appear as three nearly similar curved lines meeting at nearly equal angles. The tail is fairly long, but well whipped, or carried rather close together, and the sickles and coverts should be narrow, and tapering very gradually towards the points. Both thighs and shanks are very long, the wings fairly large, and the shoulders standing out prominently from the body even when they are closed. All the plumage is short, narrow, and hard, so much so that the breast of the cock is generally bare in the centre. This bare breast-bone is not caused by wear, and is so characteristic that when absent it is sometimes artificially produced by plucking; but when any such bare strip appears in a bird whose feathers on each side are too broad, any judge who passes it without penalty ought to feel humiliated. The plumage is also marked by great lustre when the birds are in good condition. The size is very large, especially in height, some cocks standing 28 to 30 inches in the pen; but the plumage is so scanty and close, that the body always appears rather small for its real size and weight. The comb and beetle-brows, the narrow and hard feathers, the height and length of limb, the prominent shoulders, and the "three curves", are the main points of a Malay, and so prominent that no judge can be excused for overlooking them.

The hen has the same prominent shoulders, the same type of head and neck, and the same general carriage, somewhat less pronounced. She has a peculiar habit of "playing" her tail about more than other breeds, and this point is rather valued as evidence of good blood.

In regard to colour, all varieties should have brilliant red faces and appendages, pearl, or yellow, or daw eyes, rich yellow shanks with large scales, and yellow beaks, or yellow and horn; or horn-colour is allowed with very dark plumage. The most common colour shown is a Black-breasted Red cock, with Red Wheaten or cinnamon-coloured hen, a combination which is natural, and breeds true from single mating. Pure Whites are perhaps next common, but less so than many years ago. Of late we have several times seen birds of more or less brown or ginger-breasted type. Many years since we used to see really magnificent Piles, applying the adjective to both size and colour; but this variety seems rare now. We have known it imported direct in the old days, but it is also easily bred from Whites and the Black-breasted Reds, and it is probably that really good Piles would again be popular at the present day.

Fighting

The type varies somewhat, however, as we traverse the Indian Archipelago. In the first edition of *The Illustrated Book of Poultry* was given a reproduction of a coloured drawing by a native Chinese artist of the highest class of fighting Game-cock used in the Malay peninsula, Sumatra, and the neighbourhood. This bird presented the low carriage and much of the symmetry of the Aseel, but with more full and flowing plumage. Allowing for the different style of a native artist, the breed was evidently the same as figured in old American poultry books of 1853, as the Sumatra Game, which had the same flowing tail and sweep of outline. The sickles of the Sumatra Game Fowl, carried low as they were, very nearly touched the ground, and the breed had a small and beautiful pea-comb. Other American importations were known as Javan Game and Malacca Game, all of which had very similar characteristics, but the Javan and Malaccan being larger in size. They all had a small pea-comb; and they are all reported as "dead game", beating the best English and Spanish Game then fought in the United States; one of these Eastern birds is recorded as having won no less than 75 battles against all comers.

The retired Indian officer to whom we were indebted for the drawing above alluded to, part of a collection originally intended to illustrate a work on Indian fowls and cock-fighting, supplied some interesting particulars of Eastern methods of carrying on that sport, which differ totally in many ways from those formerly practised in England as described previously. He was stationed for years in the Straits, and told us that in some districts almost every native walking about would have a cock under his arm ready for any challenger. This was specially the case in Sumatra, and at a great ceremonial cock-fight sometimes a thousand spectators would assemble. The methods of fighting were briefly as follow:-

Some birds live for years and win many matches, for generally one escapes altogether. Malay cockfighting is really much less cruel than English; a few minutes and the longest fight is over. The spurs vary in outline, some being straight, some curved, and some waved; but all have edges as sharp as razors, and are in fact like blades of penknives fastened on. This makes the fighting so quick. It takes yards and yards of soft cotton thread, wrapped round and round in all sorts of ways, to keep the spurs firm *in loco;* and this is the first art of a Malay. The *golok* (a straight spur) is generally fastened under foot, close to the ground; the crooked spur in the natural position. They take a long time to heel the birds, and lots of people (friends) look at the position, and give their advice. All this time the money is collected on the mats — piles of dollars on either side — for they are very clannish, and if one side puts down a thousand dollars, the other must do so, or no fight; that is, unless a quarrel ensue, and they fight each other. Very few English engaged in the pursuit — I did not know above half a dozen that ever did; there was some danger of rows, and few liked to have to do with it, though nothing like so bad as an English cockpit. I *once* went into the pit at Westminster, and was so disgusted with a main, I never repeated my visit. I never saw a fight at Malacca; they fight there sometimes, but it is the purely native States that make such a business of it. The Rajah of Siak, the first cockfighter of his day (1825-6), once sent a deputation to me of five boats full of officers, and about *thirty cocks,* with a pedigree to each bird: they were various colours and various names, and fine birds all. It was quite a grand ceremonial.

Many of the birds are carefully trained. I have seen a man throw down a bird and hold out one finger two or three yards off, and the bird would fly at it and strike it! The birds know their owners, and they handle them most dexterously. They are generally put out of hand on the ground by the competitors at say eight or nine yards apart; but each man seeks to put his bird down at advantage, and there is manoeuvring. The result depends much on training. Some run under and others fly high; it matters not how they meet, but meet they do, and strike home! They often meet high up in the air. I have seen — at different times, of course, and different birds — two cuts from Malay spurs, which, if they could have been done at once, and in one bird, would have quite cut the fowl in two pieces; one cut going clean through the back deep into the breast, and the other through the breast deep into the back — so keen are the edges of these deadly weapons, and so dreadful are the wounds. Generally one cock at once falls dead or next door to it, so that the other has only to give just one peck and rise, and it is over; but sometimes the dying bird lays hold of the unwounded one, and by a well-directed blow kills his assailant at once, and wins the battle. They are seldom touched after once let go, because, as I said, one is *hors de combat.*

When the Bugis come to trade in the States the betting is very heavy; and sometimes when a man loses all he has becomes desperate – In Malay language, *"meng-a-mok"* (Anglicc, "runs amuck"), and perhaps kills many. It is quite a royal affair when Bugis chiefs and Malay rajahs meet, and most intensely exciting, as they all have weapons ready for the least affront, and no man can offer another a greater insult than saying to him, *"Eteeh ber taji" (i.e.* "Duck-spurred") — the contrast is between *the duck* and, to their minds, the noblest of birds, a Game-cock! I have seen hundreds, and even thousands, of dollars lost and won on one fight of a few minutes' duration; and they go on most of the daylight after they once begin, about noon.

ASEEL

It is uncertain how long the true Aseel has been known in England. In 1871 Mr Joseph Hinton, writing upon Malays, gave the following account of some other imported fowls which he had seen:-

Last year I saw some birds brought from India by a friend. These birds he called *Game,* but in many respects they more resembled Malays. The cock's comb and gills appeared to have been cut; the shoulders were very prominent, and of extraordinary breadth for the size of the bird; the weight probably under six pounds, but the size and hardness of thighs something marvellous. The thickness of the neck was also another marked point; the hackle was scanty, and the tail drooping; whilst the general carriage was very Malay. The hens were even more Malay in character than the cocks, and their combs appeared warty. Of these birds my friend was remarkably proud. No strain could stand against them in fighting in India, and he had been offered fabulous sums for them. The hardness of these birds was something quite out of the common, and he tells me the same bird has fought four days following. The method of fighting there is a test of pluck and endurance, for they cut off the spur and bind tape over it, so that the battle is lengthened out; yet, he says, these birds would fight day after day for the time I have stated.

Aseel in England

Mr Hinton believed that these birds were probably a cross between English Game and Malays; but there can be little doubt now that they were Aseel, which were not known at that date. The details about fighting with muffled spurs are very interesting when compared with those above, by an authority who really understood Indian cock-fighting, respecting the sharp and deadly character of real combat. This latter was little test of endurance at all, but depended upon muscle and quickness; and upon that very account, as we have heard also from other sources, the muffled fighting, besides, was practised as *training,* in order to produce that hardness of muscle for which the Aseel is distinguished. For the modern introduction of the breed, however, fanciers are chiefly indebted to Mr Charles F. Montresor, who both imported it and spread the knowledge of it to the best of his ability, by exhibiting, and offering classes, and in other ways.

The following notes upon Aseel are kindly contributed by Sir Claud Alexander, Bart., of Ballochmyle, Ayrshire, whose attachment to the breed has lasted many years.

"Aseel, as their name (which is an Indian adjective meaning "highborn" or "aristocratic") denotes, are perhaps the oldest breed of domestic poultry in existence, having been kept from time immemorial by princes, and indeed by all classes in India, for fighting. How well they have been selected and bred for this purpose, will soon be apparent to anyone who takes them up; for so inborn in them is the spirit of combativeness that even tiny chickens, before they have exchanged their down for feathers, will fight to the bitter end, while the introduction of a new hen into a pen always leads to many bloody heads, and often to more serious damage in the shape of broken beaks and blinded eyes. This, from the point of view of the English exhibitor, is their greatest drawback; for even when a goodly number of promising chickens have been hatched, no amount of care will prevent some of them from being ruined for the show-pen by their brothers and sisters. Added to this, although their plump breasts and freedom from offal make them excellent table birds, they are bad layers, and the hens cannot be depended on to lay more than eight or at most a dozen eggs each. Considering all this, it is, perhaps, not to be wondered at that many who have set themselves up with a stock of this variety, have given them up in despair, and been glad to exchange the mangled remnants of their carefully collected pen for a more peaceable breed. Even as I write, my two best pullets are dead, lying slain by a sister of the same hatch; yet oddly enough Aseel show little or no inclination to fight with other breeds, though in every individual of their own race they seem to see an hereditary foe.

"Their Indian originators have not confined their efforts to cultivating the mental characteristic of their birds, but have been equally careful to develop them physically to the best advantage, selecting always those hens to breed from which were best suited in appearance to produce fighting birds, while in the cocks, survival of the fittest has been secured by the simple process of fighting them incessantly. No one who has seen and handled a good Aseel can fail to admire the skill which has produced such enormous power in so small a compass; while offal has been reduced to a minimum, and dubbing rendered unnecessary; the tiny pea-comb giving no opportunity to an adversary, and the wattles being practically non-existent. For days before the battles come off, the natives will argue and wrangle as to the prospects of their respective favourites, and in many cases before the arrangements have reached completion, the owners instead of the birds have come to blows.

"I am told by friends who have watched these fights in India that the birds in common use are of all colours, as is the case with those seen at English shows. Through the kindness, however, of Colonel Hallen, who probably knows more about Aseel than any other Englishman, I obtained some birds whose parents he had imported from the most carefully kept collections of Indian princes, and these were all either black-red or bright ginger, while a few of the hens showed faint traces of the lacing to be seen in our so-called Indian Game, which have undoubtedly been manufactured from the more ancient Aseel. Colonel Hallen informed me that no other colours were admitted in the best strains, and indeed he once expressed to me his horror at receiving from a well-known and successful English exhibitor a spangled cock of the now fashionable colour, which he promptly returned. That he subsequently accepted a black-red cock from my own yard as a change of blood, may be taken as proof that the English show bird is in all, save colour, a worthy descendent of his warlike Indian ancestor.

"The Standard now embodied in that of the Poultry Club, was drawn up some years ago at the request of several admirers of the breed, by Mr Charles F. Montresor, and those who wish to take up this interesting variety cannot do better than study it carefully. No remarks on Aseel would be complete without a reference to the great benefits also conferred on the breed in England by such careful breeders as Mr James Hutchings, Major Dunning, Mr Stawell Bryan, Mr Peele, Mr F.C. Tomkins, and Mr E. Leake, all of whose names are household words in the annals of the show pen."

The Aseel resembles the Malay somewhat in the high and prominent shoulders, drooping tail, and short and narrow feather; but the shoulders are a little less angular, and the bird has much shorter and more powerful limbs, striking one as a little low in carriage of the body. The most marked characteristic of the race is weight compared with size: taking in hand an apparently small bird, it feels "like lead" compared with any other fowl, the Indian Game coming next to it in this respect. This arises from the extraordinary *density* of muscle which has been produced by generations of severe competitive selection, and we fear it must tend to decrease as the Aseel is bred year after year without that training and selection, as a mere fowl. Rigorously to discard any approach towards "softness" either of flesh or feather, is all that can be done to guard against this tendency.

INDIAN GAME

The Indian Game known by Lewis Wright were quite different from the birds seen today. Indeed to use an analogy would be to compare the bulldog of 100 years ago with his modern equivalent. In both cases the emphasis has been placed on "more" or "heavier" bone with the result that the conformation has changed completely. Any suggestion of the birds being of the game-type has now disappeared: the longer legs bent at the hock, the upright stance and the ability to move when attacked have all disappeared.

There is no intention of decrying Indian Game, but rather to state that they have changed enormously. Selection by breeders, year after year, has resulted in birds which are broader and shorter in the leg. These are the types which win at shows and, therefore, are bred by the fancier who wants to succeed.

The two colours available in Britain are Dark Indian Game and Jubilee Indian Game. The Darks are basically black and brown with double lacing on the females. Jubilees are white with a chestnut red colour.

A brief description of the female for the Dark Indian is given below from **Understanding Indian Game** — *K.J.G. Hawkey:*

Colour: The ground colour is chestnut brown, nut brown, or mahogany brown, head, hackle and throat green glossy black, or beetle-green with a bay or chestnut centre mark; the breast commencing on the lower part of the throat and expanding into double lacing on the swell of the breast of a rich bay or chestnut, the inner or double lacing being most distinct, the belly and thighs being marked somewhat similarly and running off into a mixture of indistinct markings under the vent and swell of the thighs. The feathers of the shoulders and back are somewhat smaller, enlarging towards the tail coverts and are similarly marked with the double lacing, the markings on the wing bows and shoulders running down to the waist are the most distinct of all, with the same kind of double lacing, and often in the best specimens there is an additional mark enclosing the base of the shaft of the feather and running to a point in the second or inner lacing.

With the Jubilee Game the same description applies except there should be white in the place of black.

This breed has been familiar in Devonshire and Cornwall for at least sixty years, but has only been practically known to any extent outside of those countries since about the year 1875, being at that date often spoken of or referred to as "Cornish" Game, in recognition of its local character. For years previous to that it often received and filled classes at the local shows, and in 1870 we found a large and good collection at the Plymouth show of that year. We had at that date never seen the true Aseel, and our idea was then that the breed had probably been produced by crossing Malays with English Game. Other various accounts have been given of its origin. The late Mr Comyns leaned to the opinion that it sprang from Game and Malay "with a touch of Aseel and Indian Jungle Fowl"; and Mr Tegetmeier also believed it to be mainly Malay. It was known as Cornish "Game", because on many occasions the fowl was actually fought by the Cornish miners, being in the early days — as we know from many sources — fierce and possessed of some courage. But even at its best it was never able to stand against good English Game, being too heavy and slow, and lacking in spirit in comparison; and any fighting capacity which it ever did possess has now almost disappeared.

Origin of Indian Game

There can, however, be no doubt at all now that the true ancestor of the Indian Game fowl is the

Aseel, from which is derived the pea-comb, more moderated carriage and proportion, and more rounded form. The chief question really debatable had been, whether the Aseel had been crossed with Malay, or with the British race; and this is practically set at rest by the direct affirmation of Mr Montresor. That gentleman published a statement in *Poultry* a few years ago, to the effect that in 1846 he had been personally informed by the late General Gilbert (afterwards Sir Walter Raleigh Gilbert) how he had himself originated the breed in Cornwall, years before that, by crossing red Aseel, which he had imported direct from India, with English black-breasted Red Game of Lord Derby's strain. Taking into consideration date, and social position, and locality, and detail, this statement must be held to settle that question in the main. But from inquiries we made in various directions respecting changes which we noted with our own eyes in the birds as exhibited, there can be little doubt that some further modification of the breed took place about 1870-77, crosses being made with birds intensely black in the cocks and magnficiently glossed in both sexes, then exhibited occasionally as "Pheasant Malays". From this cross was derived a solid black breast and darker colour in the cocks, and greater richness of colour and more iridescence of the lacing in the hens; and we suspect the *double* lacing also, which we never remember to have seen before. What this "Pheasant Malay" itself really was, we are at the present date unable to say. It was certainly not Malay as otherwise shown; being smaller, with fuller tail, and with more symmetry and rounded shoulders, and it had a pea-comb. Neither was it Aseel as now shown; having too much tail, though of very narrow feathers, rather too much limb, and too upright or Malay a type of carriage. Our own impression, confirmed by every American fancier who has ever seen the breed in the United States, and whom we have been able to consult, is that these birds were probably specimens of the magnificent Sumatra Pheasant Game fowl, and that this breed has, therefore, been a third component of, and given the final "polish" to the Indian Game. Its own close relationship to the Aseel has been hinted at in discussing that bird; and upon the whole the successive mixture of strains here indicated appears the most probable pedigree of the present Indian Game.

Characteristics of Indian Game

The general appearance of the fowl is very much what might be expected from such an origin, but yet with a character of its own. The cock's head is rather broad and beetle-browed, but not nearly so much so as that of the Malay; and longer than the Malay, but not nearly so much so as the English Game. It is surmounted by a triple or pea-comb, which is apt to be rather large, and is very often dubbed. Wattles and ear-lobes are small and brilliant red, beak either yellow or horn-colour, or a mixture; eyes full and bold, varying in colour from pale yellow to red, the latter colour evidently coming from the Derby Red cross, and being strong evidence of it. The neck is of medium length, and rather arched, with short hackles, but enough to just cover the base of the neck. The body should be very thick and compact, large and broad round the shoulders and tapering towards the tail, with a wide and deep but well-rounded breast and tolerably flat back, with the shoulders standing out well and prominently, but not so as to cause a hollow back; neither must the bird be flat-sided. The wings are rather short, and carried close, with well-rounded points closely tucked in. The thighs and shanks are stout and only medium length, not nearly so long as in the Malay, the shanks being rich yellow or orange, the feet well spread and flat, with the back toe well down and almost flat on the ground. The tail is medium length, with narrow sickles and coverts, very hard, carried drooping. The whole carriage is very upright, with high shoulders and lower stern, the back sloping; altogether with much of the Malay character, but much tempered down and differently proportioned. The plumage is throughout short, hard, close, and extremely lustrous. The breast, under-parts, and tail are rich, glossy, green-black. The head is the same green-black, but the hackle lower down is mingled or streaked with rich bay or chestnut, as are the saddle hackles, the shafts of the feathers being deep crimson-brown. The wing-bows are a somewhat similar mingling of deep bay or chestnut with green-black; the wing-bar green-black of the most lustrous kind; the secondaries deep bay on the outer web, and black on the inner webs and ends of feathers, forming a chestnut triangular wing-bay.

The general characteristics of the hen are similar, allowing for sex, but her colour is different, and almost unique among the black-breasted Reds. Her head and upper hackles commence also as rich green, but lower down the centre of the feather becomes chestnut, with only a green border. The body generally is of a very rich bay or chestnut ground-colour, each feather laced or edged with beetle-green, with such iridescence as causes the lacing to look as if embossed or raised above the surface of the feather. At the throat and upper parts of the breast this lacing is very often single, but lower down, as the feathers get larger, and on the back, and on the wing-bows, there should be a second or inner lacing. I have seen photographs of some feathers of a beautiful hen, sent for the purpose by the writer of the notes which show the type of this double-lacing admirably. The feathers on the wing-bows are generally amongst the best for marking, and the coverts or bars should especially be well and boldly laced. Sometimes a *third* black centre-mark may be found in the larger feathers, inside the second lacing, but this is not usual in England. In America, however, where the Indian Game has become very popular since 1890, many breeders try to cultivate a still further or "triple" lacing, and the American Standard, till a few years ago, actually stipulated for "two or more" lacings on each feather. This is the principal difference between American and English ideas concerning these fowls, for in England even double-lacing is by no means apparent upon all feathers, though acknowledged and held desirable. To lay stress upon yet further lacing, must not only inevitably lead to the mating-up of special pullet-breeding pens, but would probably destroy that grand and "embossed" character of the marking which is so great a beauty; and on both these grounds it is a matter for congratulation that the last edition of the American Standard has omitted the words "or more", leaving only double-lacing as its description of the hen's plumage.

MINORCAS

This breed in all probability came to England from the island whose name it bears, and more than one importation appears to have taken place. The late Mr Leworthy of Barnstaple, who had bred it since about 1830, told us that several lots had come from Minorca, and that a friend and townsman of his, a Mr Willis, had been familiar with similar birds in the island itself. The Rev. Thomas Cox, of Castle Cary, was personally informed by Sir Thomas Acland that his father, the previous baronet, brought birds from Minorca direct in 1834 or 1835, from which a strain had been bred at Holnicote for many years, and distributed through the neighbourhood; and the Acland family believed that the introduction of the fowl from the West of England was mainly due to this importation. Many strains probably did descend from the Holnicote blood; but there is strong evidence that even before that Minorcas had been known in the West of England, and at the middle of the nineteenth century there were obvious *differences* between certain strains which bear out the supposition of several distinct importations. Mr Leworthy, for instance, gave as his average weights 5½ lbs and 4½ lbs, whilst other strains were considerably heavier even at that time.

This fine race had been known and valued in the West of England, from Cornwall up to as high as Bristol, for very many years before it attracted any attention amongst breeders generally. For its localisation there, the only reason that can be given is a somewhat special intercourse with Spain which has also left other traces here and there, and of which some slight indication is given in Kingsley's *Westward Ho!* The first brood of chicks we ever hatched, about 1850, were from Minorca eggs; and knowing the fowls thus from childhood, and their qualities, it was always a mystery to us that they should be so long confined to one corner of Great Britain. It is perhaps noteworthy that many of the people who kept them at that early date, called them "Black Spanish", and that throughout the West, when the "Black Spanish" was spoken of, it was generally the Minorca that was really meant. Many of the finest we ever saw were in the possession of poor men, who kept them for their eggs, which they sold new-laid; and though they never exhibited, were proud of their fowls, and in some cases refused a guinea for a favourite bird. In spite of this solid merit, and its striking appearance, however, the breed was very slow to make way out of its own corner. Our persistent advocacy of its merits had little effect for some years; but all of a sudden it began to "move" and since then its progress has been rapid. In 1883, soon after the movement had begun, there were at the Crystal Palace show only two classes and thirty-two entries; but in 1888 the newly formed Minorca Club held its first show at that great gathering, when there were six classes and 144 entries. There are few shows now which do not give Minorca classes.

Characteristics of Minorcas

Broadly speaking, the Minorca may be said to resemble the Spanish fowl without its white face, and with a much smaller white ear-lobe; but there are perceptible other differences in detail, the body being more massive and compact, the legs shorter, and the comb of different texture. The head of the cock should be large and broad, without which the comb cannot be carried firmly; beak dark horn-colour; eye full and dark; comb single, upright, and straight; large, but not extending beyond front of the beak, and falling well back behind, but not touching the hackle; it should have a few bold serrations arranged in a nice arch, and is preferred rather rough in texture; it must be a rich bright red. The wattles are long and full, free from folds also rich red, as must be the face, the latter having as few coarse hairs as possible, and perfectly free from white. The ear-lobe should be smooth and flat, almond in shape, and of colour and texture like fine white kid. The neck is rather long and curved, with full hackle; body compact and somewhat square, broad at shoulders, with full rounded breast, and the back broad and rather long; tail full, and carried well back, not upright. The legs and shanks are medium in length, the latter a very dark slate colour. The plumage is glossy black all over, but especially in the hackles, where gloss should be very great. In the hen the ear-lobe, though still oblong in shape, is rather more rounded than in the cock. and the comb arches over on one side of the face, but should do so in a manner not to obstruct the sight and not be so large as to do this. The tail should be carried well back. The carriage is sprightly an graceful in both sexes: the average weight about 7 lbs and 6 lbs for adults, and 6 lbs and 5 lbs fo cockerels and pullets.

Exhibiting Minorcas

Minorca cockerels want all the run and exercise you can give them. They can be kept together until their sickle feathers begin to bend. Then the best should be put in separate runs, as they are great tyrants, and damage each other's plumage. They must be well watched, as some will fret themselves, and once a cockerel goes back he very probably is spoilt. Give him an old hen or pullet a few days; this gives him pluck, and makes a man, as it were, of him. When about to show him, train well by taking him off his perch of an evening and placing him in a show pen, and using the judging stick around him. Repeat this a few days, and you will get him to stand up well and show himself to the best advantage. Also place him in a hamper a few times for an hour or so. Many a prize has been lost and tails broken for want of a little training before sending to a show. The best thing to clean combs and wattles is soap, a stiff nail brush, and ice-cold water. If this will not make them red and healthy, keep them home. Pullets can be shut up more closely, and shifted from run to run to stay maturity. When they shoot their combs, it is wise to bring out the colour of the plumage, lobes, and bloom on the comb. The great secret of success in the show pen is condition. It is not the slightest use in severe competition to send Minorcas unless they are at their very best. Pullets look charming a few days before laying. Cockerels often get so excited at shows that they never regain their appearance lost whilst in the show pen. A great deal of adverse criticism is from not realising that the bird may have quite altered from one show to another.

A great deal has of late been written on lobes, but the Minorca Club's standard has been adopted by the Poultry Club and by others. The lobe in shape is as the Valencia almond. The Standard states: 'Almond shaped, fairly large'. This does not mean either too large or too small. It is impossible to lay down the size in every bird, but it should be a lobe that helps make the bird well balanced in head points. For measurements sent me by several noted breeders I take a lobe of a full-grown cockerel to be at the outside depth 2¼ inches; in width at the widest part, just below the top, 1⅛ inches; at the base, 3/8 of an inch. The pullet: 1¼ inches in depth, 1 inch at its widest part, tapering to correspond with the cockerel. This on paper looks large, yet on the bird is not so; this very size has been reported on by judges as 'could do with more lobe', and in the pullet as 'fair lobe'. The main point at issue is really *shape.* Several well known winners have round Hamburgh lobes. Others, again, are wider at the bottom than at the top; either of these looks immense and out of place.

Exhibitors and judges should understand that the commercial value of a bird is very little. To be worth £5 to £30 they should bear criticism from beak to tail. Exhibitors that are real fanciers expect judges to handle well, to see that the face is really red and likely to remain so, that the comb is upright, free from false spikes or thumb marks. In hens or pullets, especially note if combs are evenly serrated, plumage is genuine to wing ends, and if good body and carriage. The shape and condition of the lobe is an important point, but a tinge of red should not overthrow an otherwise good bird. Some ridicule has been cast on the point of white in face being a 'fatal defect', and yet prizes having been awarded to birds showing this defect. No judge can always decide by the standard, as in many classes none approach the standard; but all should be guided by it. A bird showing this hated defect has a defect that is fatal to any rank as a standard Minorca. This to me is the common-sense meaning, and was so meant when the scale of points was drawn up. While this defect is really taken at its real value, I have no fear of the red-faced Minorca ever disgracing the old fanciers who so manfully stood by it.

Figure 2.17 Black Minorcas

POLISH (POLANDS)

Polish fowls were formerly called Polands,* but the latter name was gradually superseded from a conviction that the birds had no real connection with Poland, and that this was a mere colloquial corruption of their "polled" or crested character. This is still the most probable hypothesis; though the recent discovery of races of fowls with crests and beards and whiskers throughout South Russia, has perhaps added somewhat more plausibility to a possible geographical origin of the name, than existed some years ago.

The chief outward characteristic of all Polish fowls is the crest; but this is connected with a craniological peculiarity still more remarkable and distinct, though not so evident to mere observation. It consists of a spherical protuberance at the top of the skull, generally pierced by apertures which are only covered by skin, and whose size is in proportion to that of the crest, so that the best crested birds can be known as soon as hatched, from the size of this protuberance alone. Excess in one part being often connected with defect in some other, as Mr Darwin pointed out, the skulls with this peculiarity usually show a chasm in the inter-maxillary bones, which in other fowls support the roof of the nostrils; owing to which deficiency in bony support the nostrils of all heavily crested fowls appear flattened and depressed, and yet cavernous in character.

Besides the crest, the majority of the Polish varieties now bred are furnished with abundant beards, and side-muffs or whiskers covering the cheeks, while in such birds the wattles have entirely or almost disappeared. These features have however been subject to changes of fashion and breeding in the history of the fowl. Not only is the present white-crested Black variety still wattled and beardless, but before the era of poultry-shows the Spangled breeds in England were the same, and the late Mr John Baily has left it on record that the first bearded specimens, then called by him and other dealers "Muffeties", were not regarded as true. The late Mr Baker is believed to have been the first importer of bearded specimens, and supported them; and this type finally prevailed in all but the white-crested breed. There is one more peculiarity of the race, in a two-horned or double character of the comb. This is sought as small as possible, and is in most specimens almost invisible; but however rudimentary it may be, the double character can be discerned; and in allied breeds like Crêves and Houdans the double development becomes very marked. This general tendency to development of bifurcation in comb, protuberance in skull, beards, and whiskers, in combination with large crest, is remarkable as the type of a race which has profoundly impressed the poultry of France, Holland, Germany, and Russia.

The varieties of Polish generally recognised at present are white-crested Blacks, Gold and Silver Spangled, and Chamois or Buff Laced. Others can only receive mention.

White-crested Black Polish

The White-crested Black Polish is tolerably uniform in size, the finer specimens usually reaching about 6 lbs in the cock and 5 lbs in the hen. Our own opinion is that it is usually the most delicate of the varieties; but Mr Peter Unsworth, who bred it many years, reported it as hardy even in a wet and damp situation. The body is neat and compact, with fine bones and a flowing tail in the cock; and the carriage, as in all Polish, may be best described as suggestive of foppishness in the cock, and inquisitiveness in the hen. The plumage of the body is glossy black, of the crest pure white, except that there are always a few black feathers, the fewer the better, in the front over the nostrils, which it is a pity are not mentioned in the Standard, as they are always there unless trimmed away. The face is smooth and red, wattles rather long and red, ear-lobes small, round, and white, beak dark, legs dark blue or nearly black. The comb should be practically absent, but on close inspection two very small horns can generally be discerned.

In breeding this variety, the chief point is to get birds with as good crests as possible on both sides, as regards both size, shape, and colour, in which is included snowiness of the white, and the fewest black feathers. If some choice has to be made of defect, a good large crest in the cock is of more importance than in the hen. A single mating will breed both sexes alike good, if the parents are satisfactory; though of course if a good-crested cock is mated with hens not so good, more well-crested cockerels than pullets are likely to be produced.

The same remarks apply to the White-crested Blue Polish.

***Editorial Note: Fashions change! They are now called Polands again.**

Gold and Silver Spangled Polish

Spangled Polish are bred in two colours, Silver and Golden, in which alone their difference consists. The Gold would appear the oldest and strongest strain, as a great many instances are recorded of Silvers breeding Golden specimens, while we can remember hardly any case of Golds breeding Silvers. Golds and Silvers have been crossed by many breeders at different times, the produce being almost always distinct Gold or Silver, without weakening of either colour. Notwithstanding this evident community of blood, the Silvers exhibited have as a rule been larger than the Golden.

The crests of both Spangled varieties are relatively larger than in the White-crested Black, and that of the cock generally spreads more open; but it is desired as free from any hollowness or "pancake" formation, and as full and round on top, as possible, rising well up in front, and falling down towards sides and back with no split or division. It has already been intimated that beardless Spangles were first known in England, and these were more on a par with the Blacks in point of crest; but with the bearded race came great improvement in this respect. Taking the Silver for example, and simply substituting a golden ground for silver in the other breed, the crest-feathers of the cock somewhat resemble hackles in structure, and are black at the base, white in the middle part, and tipped with black at the ends, the white however increasing with age. That of the hen should be filled up as round as a ball, and also changes with her age. The first year it is black in the centre and edged with white, the width of this edging differing with the heaviness of marking elsewhere; but after the moult the centre becomes white, with a heavy lacing of black and later this may be edged with white, the crest getting lighter, with perhaps some quite white feathers, as the bird gets older.

In regard to the rest of the plumage there have been considerable changes. The birds first shown were spangled with round *spangles* all over, except on the wings, which were always more or less laced; the remains of this marking being seen in the hackles today. But this spangling was never so perfect as in Hamburghs, and the superior beauty of laced marking soon brought it to the front, where it has remained the accepted standard ever since. The beard of the cock should be thickly laced, or very dark: his hackles and saddle feathers tipped with black; the shoulders more heavily spotted, but with a notch showing some little approach to lacing; the feathers of the wing-bars almost perfectly laced, and the breast a broader lacing of more crescentic character, but still going round the feather.

The beard of the hen should be full, and well marked with black, but is not so very dark as in the cock. The neck-hackle should be well tipped with black, the tipping being a kind of semi-lacing: the breast heavily laced, thicker at the tips of the feathers, but less crescentic in character than in the cock; the wing-bow, bars, and tail-coverts rather heavily but more evenly laced, though the marking is almost always rather wider at the tips of the feathers. Her secondaries are very evenly laced, and the tail-feathers should be well edged, with a thicker crescentic spangle at the tip, and as clear as possible, though here also there is generally a little pepper or grey.

The ground-colour of Silvers should of course be as silvery as possible, and demands the same careful selection in breeding, and care to preserve it, as other white breeds. In Golds the colour of the cocks is a deep golden bay on the breast, and more reddish bay above; that of a hen a golden bay. The legs of both varieties are a dark or slaty blue, and the beaks dark horn or dirty blue.

Breeding Spangled Polish

In regard to the breeding of Spangled Polish, we have to consider the probable causes of the change or deterioration in marking — whichever it be termed — which has undoubtedly taken place during the last fifteen years. At show after show which we have visited, we have found the Spangled Polish (when there were any at all) very fine in crest, except for too much openess of centre in the cocks, but with obvious return to very poor spangling on the breast, instead of the true lacing which is recognised as the correct type. We could understand deliberate preference for really good round spangling; but the marking here referred to has not been that; it has been very poor, irregular, imperfect tipping of the feathers, not even a regular crescentic spangling, and every breeder who we have asked, has admitted it as a fault. No doubt this has partly been due to lack of breeders, and consequently of blood, and of any encouragement at poultry shows. But the reason is far more to be found in the too great disregard of marking, and practically matching birds together *solely for crest.* This may have been a necessity, since crest has been the chief point in judging. But good lacing is never an easy marking to breed; and if it be really neglected in competition with some other point, it must inevitably suffer, as it has done here. The

Figure 2.18 Polands

need of care, and the evil which may follow any return to a bad spangling, is curiously illustrated by an experience related in a letter to us from the Rev. G.J. Horner. The Golden Polish that came to him from his father, the late Dr. Horner, were spangled, Dr Horner preferring that style. These birds soon gave him offspring *devoid of all marking,* so that he had some difficulty in breeding the stock up again. Such a result shows clearly the necessity for keeping up a really good and correctly laced marking.

Good crests must of course be selected, especially in the cocks, and crest may also be cultivated to some extent by breeding only from old birds on both sides, a plan pursued by the late Mr Sylvester. But if Spangled Polish are ever to reach again even the standard of marking formerly attained, more care *must* be bestowed upon lacing in comparison with crest; the standard of perfection being the Chamois variety next mentioned, which in accuracy and beauty of marking is far superior. Birds must be selected whose lacing is sharp and cleanly cut, and as even as possible all round the feather; and in particular, cocks whose breast-feathers do not run out of lacing up the sides. Slightly dark markings are generally to be preferred, as in other cases, except that birds with the narrowest tipping in proportion to the lacing at the sides of the feather, should be specially valued. As a rule a cock with rather grey or peppered tail, breeds better laced chickens than a white-tailed one. There is no necessity for mating up two pens. While old birds on both sides generally breed the largest crests, a cockerel very often breeds the largest and most vigorous chickens.

Newly hatched chickens of the Spangled breeds are a smudgy grey or smudgy brown respectively, the darker ones being usually the best marked when moulted out. The first or chicken feathers are very indistinct and patchy, and it is only in the full plumage that the character and beauty of the marking can be seen. Except in the gradual appearance of more white in the crests, the plumage generally improves with age for at least several years.

Chamois or Buff Laced Polish

Another most beautiful variety, obviously allied to the Golden in the same way as Piles are allied to Black-breasted Red Game, and probably produced originally, though generations ago, by crossing Golden Spangled with white, is known as the Chamois or Buff Laced Polish, in which the black lacing is replaced by white, but of much greater sharpness and perfection. In regard to the points and breeding of this variety we need add nothing to the following article, kindly contributed by Mr R. Gordon, Cheviot Cottage, Leven, N.B., but we would call special attention to what is there stated regarding its hardiness. This is a striking proof of the difference between strong foreign blood, nourished by stock widely prevalent, and some present worn-out strains of Polish. The beauty and perfection of the lacing is also a proof that care for *all* the points, instead of breeding for crest alone, produces the best result in the end, even from a fancier's point of view.

"All varieties of Polish are beautiful, but the Buff-laces are truly visions of loveliness. Imagine the cock, a noble upstanding bird, crowned with a voluminous but symmetrical crest, and with ample muffling on cheeks and throat; a well-curved neck clad with lustrous plumage of orange-buff hue; back and wing-bow a shade deeper in colour, and saddle matching the hackles of the neck. Side hangers and tail rich buff, each feather sharply margined with white. The whole plumage of the hen is one uniform shade of orange-buff, every feather from crest to tail being laced with white. In the cock the crest is solid buff, but in the hen is fully laced, and is always at its best the first year. Comb and wattles are almost rudimentary in both sexes. Legs are clear blue, and beaks of a light skin colour. Add to this a sprightly gaiety of movement which seems to characterise these fowls when at liberty, and it can well be imagined that when disporting themselves on a green-sward a picture of surpassing loveliness is presented to the beholder."

Other Varieties

Many other varieties of Polish have been seen at different times, and some of them may still be found on the Continent. The *White* — all white — variety is a large and fine bearded race of fowls, but we have seen none in England since about 1880. The *Black-crested White* was said to be even larger, and probably the largest of all; this was also bearded and heavily crested, but is believed to be now extinct everywhere, as many inquiries on the Continent have failed to bring to light any survivors. The colour has been approached by several manufacturers, but the fowls thus produced were far beneath the size and character reported of the old breed. *Black* Polish have been shown in England years ago, and a few years since were reported at a Paris show: these were beardless, and rather small. The *White-crested Blue* is a recent Continental importation, though it was bred in England forty years ago, and is obviously connected with the Black variety. *Cuckoos* have been shown several times, but are not pleasing: abroad they are somewhat more often seen. The French had a variety they call *Ermine,* which is a white picked out with black very much after the colour of a Light Brahma: this colour ought to look very attractive when in good condition. *Buff* is another Continental variety not particularly rare. The original *Poultry Book* of 1853 also mentions a black and white speckled breed, and a grey or grizzled variety with heavy crests or beards, and in plumage resembling that of Silver-pencilled Hamburghs, but rather less clear than in the latter — probably very like that of the Campine. Neither of these last has been seen by English eyes for many years.

ORPINGTONS

There are seven varieties of Orpingtons, and I will name them in order of introduction to the public: Black, Buff, White, Jubilee, Spangled, Cuckoo and Blue. There are also rose-comb specimens of the Black, Buff and White varieties. The first two are hardly ever seen now, but there are a few fanciers who breed the rose-comb White Orpington, which seems to be largely developed from the White Wyandotte, as the original single comb White Orpingtons never produced any rose-comb specimens.

Black Orpingtons

The one to be first described, and the original Orpington, is so largely a Langshan that we should have included it in the earlier chapter if it had stood alone. It was originated and pushed by the late Mr W. Cook, then living at Orpington, in Kent, from which Kentish town he took the name. He stated that the method of production employed in regard to the single-combed Orpington was to cross a large Minorca cock with black sports from Plymouth Rocks; pullets of this cross being then mated with clean-legged Langshan cockerels, and the produce carefully bred to a deep-bodied and short-legged type. The result was a black fowl with the green gloss of the Langshan, but with clean legs, of the plumper make, with white skin and meat and a well-shaped carcass, and which is an excellent winter layer of brown eggs. The weakest point of the Orpington is that the eggs are not so large as might be expected from the size of the fowl; still they are, in single-combed strains, of a fair average size. Mr Cook also produced a rose-combed Orpington from the rose-combed Langshans mentioned in the earlier chapter, which had the same general qualities, but the curious difference, which we are unable to explain unless from some individuality of the rose-combed Langshans employed, that the eggs are smaller than from the single-combed. Owing probably to this difference, the rose-combed Black Orpington has never become generally popular.

There is no doubt that some original Black Orpingtons were produced as stated; but there is as little doubt that the breed has since considerably changed in two distinct directions. As stated in our next chapter, there is little question that one of the components of the Plymouth Rock was the Black Java fowl; and, as stated in the preceding, it is equally obvious that this Black Java has much in common with the Langshan, however that fact be interpreted. This darker and more typical component in the Asiatic blood had thus a double prepotency, and its predominance over the more Shanghai components would be intensified by breeding for clean instead of feathered shanks. This doubly strong element therefore rapidly overpowered the Minorca element, and the Orpingtons quickly became to all intents and purposes clean-legged Langshans, taking the place of that shorter-legged, symmetrical type once popular, but subsequently discarded by the personal feeling of Langshan breeders. In addition to this tendency, in the early days of the breed it is known that clean-legged pure Langshans, from perfectly orthodox sources, were sold to Orpington exhibitors, and appeared immediately in exhibition pens, as well as being used for breeding with their stock. This still further strengthened and hastened the reversion to Langshan type, which was then so pronounced that at many shows only one class for "Langshans or Orpington" (or the converse) was provided for the two breeds. The index of this change has lain chiefly in the size of the eggs, which has somewhat lessened since the Minorca element lost power; and in the colour of the eyes, which was often red while any foreign element remained, but has now almost everywhere reverted to the Langshan brown or black.

There has been, however, quite another change, a black Orpington of practically new blood coming

Figure 2.19 Black Orpingtons

upon the scene in 1891, when the late Mr Joseph Partington exhibited at the Dairy Show in October two cockerels and two pullets, which secured first and second prizes in each class, three of the four birds being immediately sold at £30 each; not withstanding which, at the Palace Show a few weeks later he brought out fresh birds of each sex that beat these previous winners. These birds were of a size that had never before been seen, creating quite a sensation and considerable curiosity. Mr Partington assures us that these Orpingtons also were cross-made birds, but had none whatever of Mr Cook's original strain in them at all, and that he had deliberately started with the idea of breeding himself something in the same line, but more striking and handsome. They were very large, and of splendid colour, with massive shape, and all had dark eyes. These points made them invincible in the show pen, and from this new strain, combined with the original, are descended the bulk of the winners of the present day. Many of the new strain displayed so much more fluff than former Black Orpingtons, that we cannot help thinking large females of either Black, White, or perhaps even Buff Cochin may have been employed with Langshan males.

Writing on the Black Orpington, Mr W. Richardson says:- "The first variety of the Orpington fowl to appear was the Black single comb and rose-comb. The birds were identical but for their combs, and the single comb variety was always much more popular.

"The Black Orpington is our handsomest black fowl, and to see the classes at the classical shows is a treat to a lover of black poultry. They are, as a rule, good layers of a nice brown egg, and, except for their black legs, a first-class table bird.

"The colour of the cock should be black, with a rich beetle-green sheen on all his feathers, free from bronze purple, purple barring or red or white in neck or saddle hackles. He should be massive, short, broad and deep in body, showing a wide U-shaped curve on the back and underneath, and stand on short, straight legs, black in young birds and dark slate colour in old ones. There should be no red or yellow in feet or legs, and his toes should be well spread out and straight with white toe-nails. The comb, wattles, and ear-lobes should be medium in size and of a fine deep coral red, the comb should be straight, evenly serrated, and set firmly on his head. His eye should be dark brown or almost black, the darker the better, and his beak should also be black or very dark horn colour. His plumage should be silky in texture and very abundant, and his tail and hackles should be flowing, but his tail should not be long nor the feathers too stiff.

"The hen's colour should be the same as that of the cock. She should look as massive as possible, her tail should be short, and she should have a nice wide rising cushion like the Brahma, rising gradually to the true tail feathers, as distinguished from the round ball cushion of the Cochin, which is very objectionable. The curve of her body should come straight down from her head round the breast and passing under and up to her tail in a perfect wide U shape, the same as the curve on her back from head to tail.

"Some birds show a keel which is not required, and it rather spoils their outline and appearance. The weight of the cocks should be from 9 lb to 11 lb, and that of the hens 7 lb to 10 lb. Equally good cockerels and pullets can be bred from the same pen, which saves double mating, though some strains seem to produce better cockerels and others better pullets."

Breeding Black Orpingtons

Colour should be bred for as in the Langshan, but the crimson between the toes is not required. Particular attention should be given to preserving the correct shape, with a broad and deep breast, the whole body looking massive and solid, and set rather low. Excessive fluff should be avoided, as tending to decrease laying, and being often accompanied by thicker skin; too small combs, also, are apt to be signs of diminished egg-production. We have seen one or two specimens distinctly *keeled,* almost like some exhibition ducks, and this ought certainly to be deprecated. It is probably due to carelessness of these points that statements have lately appeared to the effect that some strains of the Black Orpington have not kept up its reputation as a good layer. Both abundance and size of eggs would, however, quickly respond to selection for these qualities, in the manner insisted upon in former chapters of this work.

Buff Orpingtons

The first pair of Buff Orpingtons ever shown as such were exhibited by the late Mr W. Cook at the

Dairy Show of October, 1894, when Mr Cook drew our special attention to them, and made the same statement which has been made on many occasions, that they were produced by mating a Golden-spangled Hamburgh with a coloured Dorking hen, pullets from the produce being mated with a Bluff Cochin cock, the main characteristic of the birds being the combination of buff plumage, with *white* legs and feet. We remarked, on this earliest possible occasion, that a fowl with such points might probably prove both valuable and popular; but that there was grave objection to calling them Orpingtons, since he has already appropriated that name to another fowl, which had, according to his own account, not one single element in common. Such nomenclature would not have been allowed by the Poultry Association of America, and objection to it was widely expressed by the most prominent authorities in England, with scarcely an exception; the already existing Orpington Club also protested against the same name being given to another fowl which had not in it one atom of the same constituents as theirs. A considerable amount of discussion took place later, emphasised by the fact that precisely similar fowls were exhibited under another name at the Smithfield Club Show of dead poultry. Owing largely to this latter circumstance, the question was finally brought before the Poultry Club, who decided that it was then too late to interfere; but intimated that such a case would not again be allowed to pass unnoticed, and in this way it is to be hoped that the circumstances may have produced a more definite understanding concerning such matters in the poultry world.

The actual origin of the breed must be questioned, as well as its present name. There is no reason to doubt that Mr Cook really did breed birds as stated, and that these have been sold as Buff Orpingtons, or that to his persistent advertising and pushing it the popularity of the fowl was mainly due; the latter fact proving that capital may be employed as successfully in floating a variety of poultry as in founding a new journal. But evidence was published simply overwhelming in amount, to the effect that the stock about the country is mainly derived from breeding up to points a gradually formed and popular local amalgam of Buff Cochin and Dorking, which has long been known in Kent, Surrey, and Sussex, and still more so in Lincolnshire; whence it has been called, in market parlance, the "Lincolnshire Buff", forming a large portion of the "Boston" fowls sent to the London market.

To sum up a controversy given at some length in a previous edition, there is no reason to doubt that Mr Cook bred birds as alleged, and it is well known that some of the early Buff Orpingtons sent out bred very much as might be expected from such heterogeneous crossing. There is an abundance of evidence that all breeders who took up the new breed found plenty of work to do in it, and that some of them selected simply the best specimens they could find, wherever they could find them, in Surrey, or Lincolnshire, or anywhere. That birds were bought in the latter county of people who had bred nothing else for a quarter of a century and were shown directly as Buff Orpingtons, and used by Buff Orpington breeders, is quite certain; and various successful strains have no doubt had different local origins, which accounts for the fact stated by Mr Richardson presently, of the evil results found to follow from crossing these different strains. It is noteworthy that none of the early show specimens had the shape of the Black Orpington, all being higher on the leg, longer in the back, and less massive in the body; but breeders have, as we write in 1911, by paying attention to make and shape, overcome these defects.

The merits and utility of the breed stand apart from its origin and name. Those who objected to the latter were accused of making a "virulent attack upon the breed", but without, as far as we know, any foundation. The fowl itself was recognised by nearly all as a most valuable one, endorsed already by the long experience of the Lincolnshire breeders as a first-class breed for the market; and speedily found, as soon as kept alive for other purposes than market, to be a most hardy bird and prolific layer. Putting aside some of the claims advanced to which there are reasons for taking exception, the late Mr Cook may be given full credit for "booming" and making known in other than poultry-fattening circles, what is recognised as one of the most attractive and useful of all classes of poultry, combining the beautiful and popular buff colour, with admirable table and laying qualities. It is probable that the Buff Orpington, as now known, comes as near to the ideal of an all-round, general purpose fowl as is humanly possible; and it is not a small service to have made such a bird popular amongst breeders generally.

White Orpingtons

For the following notes we are also indebted to Mr W. Richardson:-

"I showed my first White Orpingtons as Albions at the Dairy Show, I think, in 1900, and the only White Orpingtons that were known before that time were some rose-comb birds which were evidently a cross of White Dorkings and White Leghorns, showing little of the true Orpington characteristics. The birds I showed were pure sports from Buff Orpingtons, and, of course, were similar in type except for their white colour, and I procured these four white sports as chickens from the same farm in Lincolnshire whereon I saw the Buff Orpingtons. I had myself bred a white chicken (a pullet) from my buffs the year before, but sold it, as single-comb White Orpingtons had not then been heard of. I agreed to change the name from 'Albion' to 'Orpington', as several fanciers who had taken them up thought they would go better as such, especially as they were sports from buffs.

"The birds I had at first were rather small, but of good type for those days, and perfectly white, and my stock today are directly descended from them without any other cross. I once crossed a White Plymouth Rock cock with two or three pullets, but the result was so disappointing that I killed all the chickens to make sure that they did not get mixed with my original whites.

"The White Orpington has one or two faults in breeding, but these are not very troublesome to overcome. One is that they throw a few yellow-legged chickens, a defect not due to cross, as the original Buff Orpington that they sported from did the same. Another defect is that some White Orpingtons come with a creamy or sappy colour in their feathers, but I find a very small proportion now in the total of chickens. Birds are also shown, especially hens and pullets, with blue legs, which is very objectionable, and I think due to some out cross. I have also seen a few specimens with green or willow legs. There is little doubt that when the single comb White Orpington was first brought out it became so popular that it was impossible to breed enough birds to supply the orders, and many people tried to produce them from various crosses. These attempts at short cuts, of course, led to many faults in the birds thus bred that my original birds never had. Others may have had odd sports from their buffs, and I have heard that there were white sports from the Black Orpington some years before the White Orpington was before the public, but I never saw any.

"One thing is important, and that is to be able to distinguish between a sappy bird and one that has been turned a bit yellow on the top by the sun or weather. Exposure of this kind only affects the hard part of the feather in the hackles of the cocks and on their wing bows and saddles and the hackles of the hens. Sap, however, shows all through the feathers to the skin. Some birds, indeed, look as if they had been dusted with yellow sulphur. It should also be remembered that white birds are easily stained by being bedded down with oak sawdust or straw, as when either of these or the birds get damp their plumage becomes discoloured, and this is very annoying as it is impossible to get out.

"The White Orpington cock should be a large massive bird of the same type as the Buff and Black, and should have clean ivory white legs with as little pink showing in them as possible. His colour should be snow white, and the feathers should be dense and of a nice silky texture. His comb wattles and ear-lobes should be bright red and medium size. The comb should be evenly serrated, straight, and free from side spikes; his eye should be red, and he should be free from sap in the feathers or a pink shade on the wing bow. Whites throw a few feathers with a little black or buff in them, but as a rule very slightly, and it is not of much consequence.

"The hen should be the same as the Black and Buff Orpingtons, massive and cobby, standing on good, stout, white legs, the head should be neat and the eye red, the worst faults to be noticed being blue in leg and white in lobe.

"The White Orpington is a marvellous layer of a nice large brown egg, surpassing, I think, the Buff in that respect, though some buffs run them very close. I have known a pen of my whites to average 206 eggs in the year. In conclusion, may I add that it has given me much pleasure to write these notes on three of the finest breeds of poultry in existence, and I hope they may be of some assistance to those fanciers who wish to breed them."

Jubilee Orpingtons

This handsome variety was introduced early in 1897, and derived its name from the Diamond Jubilee year of our late Queen Victoria. The birds are particularly handsome, with their chocolate ground colour, a black bar and white spangle at the end of each feather, and today are exhibited showing even markings of the three colours. They are cobby in build, are deep and broad, with short, white legs and four toes, all quite free from any stubs or feathers. The comb should be of medium size, firmly set upon their head, and free from any side sprigs; and the birds should have red ear-lobes and small, rounded wattles.

"During the last two or three years Jubilees have increased in size, and there is not now (1911) so marked a difference between them and the Black, White and Buff varieties. For table birds the Jubilees certainly equal any other fowl; they are quick growers and very prolific winter layers; and the cockerels are in great demand amongst farmers for turning down with their cross-bred hens, and excellent results have been obtained by such mating.

Among fanciers the Jubilee is very popular, as double mating — i.e., one pen for producing the best cockerels and another for breeding the best pullets — is not necessary, and a single pen will produce first-class specimens of either sex. The leading poultry shows provide classes for Jubilees, and wherever exhibited they attract attention by their pretty appearance and uniform shape; and they are suitable alike to a working-man, with quite a limited space, or a poultry farmer having acres at his disposal. They have, moreover, the advantage of having their interests looked after by the Jubilee Orpington Club and the Variety Orpington Club.

As to how Jubilee Orpingtons were produced is not generally known among many of the oldest breeders today, but I may state at the start that Buff Orpington pullets were selected of a chocolate or reddish-brown colour, and these were mated to the old Red Dorking, the cocks showing the most black in them being used. Many good coloured specimens were thus obtained by the first mating, and by continuous and careful selection of the progeny, the fifth toe and long back of the Dorking was stampeu out. Golden Spangled Hamburgh cocks were also used on to many of the above first-cross pullets, and here again the difficulty of the white lobe and rose comb had to be bred out.

Spangled Orpingtons

The year 1900 saw the introduction of the 'Black and White' variety of the Orpington family, known as the 'Spangled Orpington'. The writer must here explain that the word 'spangled' has quite a different meaning to that of the word 'tipped'. The latter should be understood as meaning a small half-moon of some colour different from the main or ground colour of the feather. By 'spangled' — or, as often described, 'spangling' — is meant that almost half the feather is covered with some other colour than that of the ground colour. Owing to the plumage being striking and made up of but two colours, this variety has made good headway, particularly in the exhibition world, as, besides being a handsome fowl, it is a true Orpington in shape and build, and easy to breed true to markings. The Spangled is, if anything, a larger bird than the Jubilees, being somewhat longer in breast and body, and therefore making a very fine table fowl, with pure white flesh. As all-round layers they are prolific, and hold their own well with other varieties of Orpingtons. Excellent reports have been received from breeders in Canada who have imported eggs or birds, as to their hardiness and capability of withstanding any climate.

In plumage they somewhat resemble the Ancona and Houdan, but are, of course, a much heavier and deeper bodied bird. The Spangled Orpingtons were in the first place simply sports cast off in producing the Jubilees. Many of these coming black and white were saved for at least one year to notice how they moulted out, and, in most instances, this proving satisfactory, they were equally divided, part being mated to very dark coloured Dark Dorking cocks, the others were put with Black Orpington cocks. The progeny from these two matings were carefully selected and re-mated, thus producing the present day Spangled Orpingtons. Breeders of several years' standing can, after carefully considering the various breeds used for the Jubilees and later for the Spangles, trace the very few faults that occasionally disport themselves in their present Spangled Orpingtons. The ground colour should be a beetle-green black, each feather having a fairly large spangling of white from the tip to almost a third of the way up the feather; the comb is a single one, of medium size, firm, and well set upon the head; ear-lobe red; legs short, stout, and well set apart; four toes on each foot, and leg colour in accordance with the latest standard is mottled black and white, though all white is allowed. Weight of cocks 10 lb, and in hens 9 lb.

Breeders should avoid using birds for breeding purposes, having the follow defects, viz. side spikes on comb, white in ear-lobe, black legs and feet, any straw, red or brown markings in the plumage.

Cuckoo Orpingtons

In 1907 the poultry fancy were surprised to find yet another variety of Orpingtons placed before them, and many remarks were expressed that, as the originator of the five former varieties had passed away, they did not anticipate a further addition. The experience I had gained with my father in the production of the former Orpingtons had naturally been of value, and as nothing is more interesting than making and perfecting a new variety, I worked at and introduced the Cuckoo Orpington. Classes have already been provided at the classic shows for the Cuckoo, and they are in the hands of many well known and prominent exhibitors, but I confess that my aim was not so much to make them an exhibition bird, as one possessing high-class utility qualities, combining a big, deep body on low legs, with prolific winter laying; and these merits are in themselves sufficient to make the variety popular. In appearance the 'Cuckoos' are somewhat similar to the barred Rock, without the objectionable long, yellow leg, which has given place to a short, perfectly white leg. The flesh and skin are pure white, the back is short and very broad; in fact, the Cuckoo is the blocky type so prominent in all the Orpingtons. They are extremely hardy, and lay a very large, rich brown egg.

Single mating only is required, so that even the purchase of a small pen is sufficient to build up a good foundation stock of a general, all-round useful and handsome fowl. The colour is a light, bluish-grey ground, having bars across the feathers of a darker blue-black, proportioned to the size of the feather and the same on all parts of the body. The beak, legs and feet are pure white; lobes red; comb small, firm and evenly serrated.

Defects to be avoided are white in lobe, any yellow in legs and feet, long legs, stubs or feathers on legs, side spikes, and more than four toes.

Blue Orpingtons

Quite the newest of the Orpington family is the 'Blue', which I introduced just as 1907 was drawing to a close; and I do not remember any variety that caused such a sensation. Fanciers recognised in it a breed that would make tremendous headway, and the demand far exceeded both supply and anticipations. 'Imitation is the sincerest form of flattery,' and though 'Blue' birds of a different shade and shape have been exhibited, the 'original' Blue stood its ground, and for colour and type has now set the standard to the whole world. A true Orpington in build, with wide chest, broad back, small head and tail, low, wide-set legs, they resemble the Black Orpington more than any other fowl. As a layer the Blue is second to none. It is extremely vigorous and hardy, thrives well in any climate, and in appearance and colour is both pleasing and striking.

In their production Blues have certainly proved the most difficult variety to produce that I have as yet experienced, for the Blue colour is extremely hard to fix. Nothing but pure Orpington blood was used in their production, and these consisted of the White, Cuckoo, Spangled and Black varieties, crossed and recrossed, and marked success, as admitted by the oldest and keenest breeders, has been achieved.

The colour in cocks is of a dark blue top colour, with a somewhat lighter blue breast and fluff, laced with an outer edging of a darker blue shade. In hens the chief ground colour is of a medium blue, each feather laced with a darker blue on the outside edges. The eye, legs and feet are black, ear-lobe red, comb and wattles small. Weight in cocks 10 lb, hens 9 lb.

Defects which should be carefully avoided are: Hazel or light-coloured eyes, white legs, white in lobe, pale blue body colour.

PLYMOUTH ROCKS

Origin of Plymouth Rocks

For years nothing more was heard about Plymouth Rocks; and in the *New York Poultry Bulletin,* the first American poultry periodical ever published, during its first two years such a fowl was never even named. Their first mention in that paper was about 1870; and in response to a direct inquiry of our own, we received the first direct information about them in a letter from Mr W. Simpson dated August 12, 1871, in which he states that their plumage was "Dominique" (the American term for blue barred or "cuckoo" colour), that they had been produced by crossing the native Dominique or common cuckoo fowl with Asiatics, and up to that time did not breed very true, while their eggs were all colours and sizes. Everything points to a new production about that time, from quite recent crosses, and there is not the slightest doubt that the present Plymouth Rock, of totally different colour from Dr Bennett's old creation, had its origin about this time and in this general way. The first ever seen in England were sent over by Mr W. Simpson in 1872, and took honours at Birmingham that year in the class for Any Other

Variety; since which time they have rapidly grown in numbers and popularity, all the original stock having come to us from across the Atlantic.

Further investigation in America has made it pretty clear that the modern Plymouth Rock had more than one origin, and that the claims of various breeders, such as Mr Spaulding, Mr Drake, Mr Upham, Mr Giles, and Mr Pitman, to have produced ancestors of the present birds, were all more or less well founded. Mr I.K. Felch, whose long memory makes him a good authority, has traced various crosses made by different breeders, including the following: (1) Spanish on White Cochin, top-crossed by Dominque; (2) Dominique on Buff Cochin hens; (3) White Birmingham (supposed to be an English fowl, but what, no one can tell) on Black Java, the produce coming as white, black, and Dominque, and the Dominiques alone being bred together; (4) the same produce top-crossed with Dominique; (5) Black Java and Dominique; (6) some of the above crossed with Brahma. This last cross distinguished the Drake strain; and it has been stated by some that the amalgamation or breeding together about 1869 of this Brahma-crossed strain, with that of another strain also containing Black Java blood, produced the final improvement and stamp which gave the new Plymouth Rock its growing popularity from 1870 onward.

This origin of the fowl will explain the chief difficulties in breeding barred Rocks. The colour itself is not a natural primary one, but the produce of white with either black or very dark colour. Such colours, mated together, produce as the result, on a wide average of cases, more or less of blacks, whites, mottles or splashes with the plumage of Houdans and Anconas, blues or blue duns like that of blue Langshans and Andalusians, and that bluish barred plumage known as Dominique in America and Cuckoo in England. When once produced, this last colour has however a strong tendency to permanence; and in the common native Dominique fowl of the West Indies and United States it had been preserved and bred so long as to be of a very fixed type indeed, though even in these fowls there was a constant tendency for the white or black feathers of the original components to appear, as well as the straw or red which always troubles breeders of white or black fowls. But in the barred Rocks, fresh blood of both white and black had been thrown in; and in the Black Java particularly, which all accounts agree in stating had been always used on the female side, a strain had been used which we have alredy seen is perhaps one of the most strong and prepotent now in existence. To this day that strong black blood is constantly occurring to an extent not known in any other blue barred fowl. These are almost always on the female side, though black feathers will also often appear in the male, which is however more subject to white than to black in his plumage. These facts are explained by the origin of the fowl, which, when well understood, affords valuable indication to the breeder as to his choice in certain cases.

Characteristics of Plymouth Rocks

The general characteristics of the Plymouth Rock are very much what might be expected from its origin. It is a large fowl, only slightly inferior in size to the large Asiatic breeds. The comb is single and straight, evenly serrated, much like a good Cochin comb, but preferred rather smaller, with wattles large in proportion; ear-lobes smaller, and red. The head and neck are carried upright, and not forward like the Cochin's. The body should be large and rather square, but with a deep and compact appearance, and the plumage close, with only very moderate fluff; wings moderate in length and carried close. The shanks should be moderate in length, not long nor yet too short, and set wide apart; they are clean, and bright yellow in colour. The cock's tail should be neat, and carried only moderately high, and well compacted; but we never could understand the original Standard, which said it was "smaller" than a Cochin's, as even in England, where the ideal differs widely from the American, we have never seen a bird without a tail much larger and higher than any Cochin breeder would recognise as at all proper in his variety. The tail of the hen, though small, is also considerably larger and more projecting than that of a Cochin.

Colour of Barred Rocks

The plumage is not easy to describe with exactness, and we have known two observers, both accustomed to consider their words, describe the same bird, and the very same feather from the same bird, differently, and in each case rather differently from our own idea as to its real colour. It is not so in regard to the barring; that runs straight across the feather, much like that of a pencilled Hamburgh, but considerably coarser; also the bars are not sharply edged, but the dark bar shades into the light through a small space, though they should not do so too gradually, or so as to destroy the distinctly "barred" effect. From about 1890 the bars have been bred perceptibly narrower and more numerous than formerly, though not so much so in England as in America; there is pretty obviously a happy medium, unless some day a fashion should set in for breeding an out-and-out "blue-pencilled" fowl; and beyond a certain point narrowness is not desired in either country. There is also a proper proportion between the dark bars and the light spaces, not very different from equal spacing being desired.

The real difficulty, no doubt, is how to describe the colour. Looking at the whole bird, in England especially, there is distinctly a blue dun shade or appearance about it; but when we examine a single feather, it is difficult to see any blue colour in it at all. The English Standard gives the ground as blue-white, evenly barred with bands of black of a beetle-green sheen; while the American Standard gives the ground as "greyish white", barred with "defined bars that stop short of a positive black". Yet average English feathers are certainly darker and with more approach to blue-dun in the ground colour than American feathers, while on the other hand American birds appear a distinctly brighter blue in the whole effect, than English birds. The fact is that the colour of a single feather is greatly affected by that of any surface on which it is laid; and when American feathers are laid upon a white surface, many of them appear merely black and white, each colour being a little dull, the white not quite pure and the black very slightly greyish. If any breeder will examine single feathers laid upon papers or cards of different colours, these remarks will be understood, and such facts make a really definite standard of colour very difficult to frame.

Merits of Rocks

It was as a very profitable and generally useful fowl all round, that the barred or original Plymouth Rock steadily achieved popularity in the United States, and later in England. The colour wears well and looks well, especially about a farm; the laying powers are above the average, and when cultivated reach a very high standard indeed; the meat, though not white or such as in England is considered first-rate in quality, is extremely good and juicy; and the bird makes a rapid growth which is only equalled by Dorking or Houdan crosses. The constitution is hardy, and the chickens easily reared. Until the White Wyandotte arose, no fowl was ever bred and kept so extensively as the Barred Rock was in the United States, and it probably holds the first place there still; and we have already seen that it forms the basis of a large portion of the best table fowls sent over from Ireland.

Breeding Barred Rocks

The Barred Rock plumage is not easy to breed to present exhibition standards, and as a rule requires more or less the system of double mating. In this case the necessity arises from fanciers desiring to make similar in both sexes, barring and colouring which Nature has arranged to be finer in barring and lighter in colour in the males than the females. That would apply to all cuckoo barred fowls; but in this particular case there is the added difficulty of the strong Black Java blood always tending to reappear in pullets, if birds too dark are used on either side*; some birds of this kind are inevitable if adequate colour is to be preserved, but they are getting somewhat fewer than they used to be.

White Rocks

"The White Plymouth Rocks", writes Mrs Wilkinson, "are beyond doubt a pure variety, owing their origin to sports from the Barred. These have not been so frequent in England as in America. They are believed to have been first preserved and cultivated by a Mr Frost, of Maine, about 1880, and later they were brought to about perfection by the late Mr Harry W. Graves, of Higganum, Conn., who won almost all the prizes at Madison Square for a few years before his death. In 1905 Mr Graves sold the champion male for one thousand dollars, so there is no question of the variety's popularity in America.

"It is said that Whites are easier to breed than the Barred variety. No doubt that is true to some extent, but to breed a typical bird of pure white with good orange yellow legs and beak, with face, comb, earlobes and wattles red, is difficult enough, and when attained the result is truly a beautiful bird. White birds, as a rule, have a tendency to straw colour or sun burn, but White Rocks in the best strains have almost overcome this and now rank amongst the most perfected varieties in existence, especially so in America, and also in one or two English yards. White Rocks should correspond in shape and size with Barred Rocks, and although not so popular they possess better utility qualities.

Black Rocks

"Black Plymouth Rocks are also sports from the Barred variety. It is seldom that a cockerel is thus produced, but pullets are very prevalent in some strains. A few enthusiasts have bred them exclusively and produced some handsome specimens. They are of the same shape and size as the Barred, and have a good lustrous green-black colour with grand yellow legs and beak; bright bay eyes, red face, comb, wattles and earlobes. The contrast is certainly very becoming, and, moreover, the blacks are undoubtedly very good layers and useful table fowls. They develop more quickly than any other of the Rock family, and yet have the smallest number of followers amongst breeders who breed them pure, though several breeders mate them up to their Barred Rock cockerels to improve the colour of the latter."

Buff Rocks

There is also a Buff variety of the Plymouth Rock now widely bred and exhibited, but which obviously cannot be considered any true descendant of the original. In the United States its chief component has undoubtedly been the Buff Cochin; in England some of the stock has come from America, but more owes its origin chiefly to the Buff Orpington or Lincolnshire Buff. There is little doubt that the one "breed" was in England at first largely bound up with the other, quite a number of breeders exhibiting both, and putting a bird into either class according to the colour of its shanks. Such a double refuge for produce was highly convenient in many ways, for the exhibitor; but it was not so good for either variety, in the long run, and there is little doubt that some of the difficulty about yellow shanks occurring in Buff Orpingtons, has been due to the allied Plymouth Rock blood, which has both given encouragement to yellow, and sent it back again into the white-legged strain. The only remedy for this, as already intimated, is greater insistence upon true type or form, in both breeds.

The following notes upon breeding Buff Rocks are also kindly supplied by Mrs Wilkinson, whose successes in this variety have been well known:-

"A Self-colour is not so complex as a bi- or tri-colour, and is therefore simpler to describe, but the production of either black, white, or buff self fowls is not so easy as the uninitiated would imagine. In the past there have been many heated discussions on the correct shade of a Buff Rock, but at present all is serenely peaceful, one shade being generally accepted as the ideal. This shade can best be described by comparing it to the colour of a new golden sovereign, which it exactly resembles. Of course a little latititude is allowed, especially in the stud birds, provided softness and soundness are retained — all hard bricky colours are most undesirable.

"Buff fowls should have one colour and one only, from head to tail, and from root to tip of each feather, with not the slightest suspicion of lacing round the edges of the feathers (generally found at bottom of breast and wing bars), nor yet possess any feathers which are 'mealy' — a sprinkling of tiny white spots like meal, generally clustered round the quills on the wing butts, but occasionally all over the body. These two defects, along with black or white undercolour, are infinitely more to be dreaded than a little black in tail or flights, and should never be allowed in the breeding stock. White should be entirely absent from young stock, but as quite 90 per cent of the good youngsters show white after the first moult, we are obliged to allow a little after this period, but even then, only a touch on flights and tail.

"I have described colour first, but I believe shape ought to have had that position. It has often been truly said that shape makes the breed, colour the variety; and this most particularly applies to the Plymouth Rock family. The Buff Rock should have the shape described above for the Barred variety, and not that of a moderated Cochin, like many that were exhibited about 1895 and 1896. Short round heads, huge cushions, and an abundance of fluff, are not Plymouth Rock properties, and should be carefully avoided.

"When mating this variety, select birds which comply as nearly as possible with the above description. Don't use a bird because of its surface appearance, but be satisfied that it is sound to the skin, and that it comes from a strain which has been carefully bred to type for years back.

"An article on Buffs, at the present time, would be incomplete without some reference to colour feeding, so I will conclude with a few words on that subject. A few years ago every prominent breeder of Buffs was under suspicion for colour feeding; why, I cannot say, for it has never, to my knowledge, been proved that it is possible to improve a Buff in colour by feeding. It may be possible to make a light colour darker, but even if so, I feel assured that the lovely soft tones one has become accustomed to see, will never be attained by other means than skilful breeding."

There is nothing to add to these observations, with those previously made upon breeding Buffs in general, under the heading of Buff Cochins and Orpingtons. It may perhaps be remarked that upon the whole this colour is about as easy to breed in Rocks as in any variety, the rich colour desired in the shanks harmonising and working in well with that sought in the plumage. For this reason it will be found, comparing two average classes, that the colour in the Rocks is as a whole slightly richer and sounder than in Orpingtons.

All the varieties of Plymouth Rocks have also been bred in America with *pea-combs.* Some of these are probably sports, from the original Brahma cross which has already been stated to have existed in the Drake strain; others have been avowedly produced by a fresh cross with the Brahma. The danger of frostbite in some parts of the States causes a preference for the pea-comb in itself, among some farmers, which does not exist in England; but in spite of this such varieties do not appear to extend much, and in England have never taken root at all. *Partridges* have also been shown, but such attempts to multiply varieties that can have really nothing in common with a breed of real character and value, are rather to be deprecated than encouraged.

THE RHODE ISLAND RED

It would appear, from available data, that this distinctly useful and beautiful fowl had its origin some half a century ago in a cross between the old Shanghai, the Chittagong, and the Red Malay, the males of those old breeds being mated with the female fowls already existing at Little Compton, Rhode Island, the most extensive of the group of isles situated in the famous Narragansett Bay, the foreign poultry hav-

Figure 2.20 Rose-comb Rhode Island Red

***Editorial Note: At this stage the breed was not established in the UK as a single combed bird.**

ing been undoubtedly brought to Rhode by the vessels plying to the island.

The American Standard expresses the belief that the Rhode Island Red "originated from crosses of the Asiatics, Mediterraneans, and Games," and it is stated on good authority that the brown Leghorn entered largely into the composition of later types. Be that as it may, the present-day Rhode Island Red is the result of fifty years of out-crossing, hence the difficulty still experienced to breed the fowl true to colour. That the object of the old farmers of Rhode Island was, from the commencement, to produce a fowl conforming to specified pure-breed characteristics may be dismissed as being highly improbable, since they apparently simply strove for the elimination of feathered legs, and the production of a bird having a "carcass" that found a ready sale in the Boston and other markets; and it was not until the fancier turned his attention to the fowl that it took on some resemblance of definite form.

SCOTCH DUMPIES

These fowls are of considerable antiquity in Scotland, of how great it is impossible to discover; and they have been known in England since 1852, when the late Mr John Fairlie introduced them into his yards near Newmarket. They were also called Bakies, Golaighs, and by other synonyms. About 1870 they appeared nearly extinct, and Mr Thomas Raines, of Stirling, wrote to us that he knew of only one or two people that still had them; but national feeling has recently made commendable efforts to resuscitate a breed which certainly has commendable qualities, and with such result that it has found a place in the Standard.

As a rule Dumpies have a rather large, single comb, fair-sized wattles, and red earlobes. The real characteristics lie in a long and large and deep body, carried upon *extremely* short shanks, rarely exceeding 1½ inches in length. In all the lighter colours the shanks are white; in blacks they may be dark. The plumage is found of all colours; and while single combs are most common, rose-combs are also allowed. These variations denote considerable mixture in breeding.

The Dumpy cock attains to 7 lbs or 8 lbs weight, the hen a pound less. The bird is a very good layer, and the flesh exceedingly tender and juicy, surpassing that of Dorkings in the opinion of some who have compared the two directly. They are admirable sitters, covering more eggs than their size would appear to warrant, and make good and assiduous mothers, who have the quality of generally taking readily to the chicks of other fowls. Taking it all in all, the fowl is one well worth more cultivation on both sides of the Border.

SCOTCH GREYS (SCOTS GREYS)

This is a most useful breed, which we have often wondered has not been more popular in England. It has long been known and valued in Scotland, but about thirty-five years ago seemed losing ground even there; more recently, however, it has been taken up with energy, and at many Scottish shows there are now large and good classes. It has been called the Scotch Dorking, but is entirely different in carriage and shape from that fowl, being more sprightly in form, with something of the Old English Game style about it. The comb is single and moderate in size, upright in the cock, usually falling over in the hen, the earlobe red, wattles medium in size, legs white or white mottled with black. The plumage in general resembles that of the barred Rock, but has a tendency to finer marking, and the more pronounced black and white of American Rock breeders.

The Scotch Grey is a very moderate eater and good forager, and an especially hardy fowl, especially in cold or damp situations. The flesh is as a rule more juicy than that of the Dorking, partaking more of the Houdan character. The breed was not formerly known as a very good layer in comparison with some others; but it has been found that this quality, as in other breeds, can be easily developed, and good laying strains formed. We know it to be making progress both on the Continent and in the United States. The fowl does not appear, however, well adapted for confinement in sheds or very small runs, being too restless or active, and (like other breeds of that disposition) rather apt in such circumstances to start feather-eating.

SILKIES

These peculiar fowls are described by several of the oldest naturalists, "hair like cats" being one of the expressions employed regarding their plumage. Hence they must have been known from an early date;

Figure 2.21 Scots Greys

but it is remarkable that some later authors, such as Willoughby and Ray, write of these accounts as fanciful and unworthy of belief. The soft and flossy plumage is not the only distinguishing characteristic, however. The skin is of a deep violet colour, almost black, and the periosteum, or covering of the bones, is of the same colour: hence the fowl, though really excellent eating, is rather repellent to ordinary notions upon a dish. There is a moderate crest, standing well up, and in the cock rather backward. The comb, face, and wattles are of a deep mulberry colour; the ear-lobes should be a bright or turquoise blue, though often tending to the same purplish tinge; the legs also are of a deep bluish black. The strong dark blood obviously runs through the whole fowl. The legs are slightly feathered, and have five toes. The general shape, in some respects, resembles that of the Cochin, with ample cushion or saddle, and short tail, but without any of the latter breed's heaviness of carriage. The size has varied considerably: formerly cocks scaled 4 lbs and hens 3 lbs; but of recent years they seem to have become smaller, and the name itself is spelt "Silkie" by true believers.

The most obvious point is, of course, the peculiarity of plumage. In fowls generally the stem of the feather is strong, and from it proceed fibres which are stiff and elastic, and furnished with fibrils differently arranged on the forward and backward sides, so as to interlock and form the "vane" of the feather. In the Silkie fowl's plumage the stem is thin and weak, and the fibres weak and non-elastic, with rudimentary hair-like fibrils which have no holding power and no locking arrangement. The result is the loose and flossy character shown in Fig. 2-22, which is a body feather from a Silkie hen. Ordinary Cochin plumage is what one might almost call half-way towards this silky character; and it is not surprising, therefore, that the Silkie should present much Cochin type, and that the Emu Cochin should be the one breed which should sometimes exhibit the silky type of plumage.

Figure 2.22 Feather from Silkie Hen

No one has probably known and bred Silkies so long as the Rev. R.S. Woodgate, of Pembury Hall, Kent, who in 1900 contributed the following notes, and who was a prominent exhibitor even before the first edition of the *Illustrated Book of Poultry* was published.

"I cannot but remark how pleased I am to write these notes, as I did in the original edition long ago of Mr Wright's *Illustrated Book of Poultry*. I have been acquainted with this most interesting breed for forty years. A Captain Finch brought home a pair about 1860, which, to my juvenile recollection, were as beautiful as any now on view. I have since endeavoured to find out where they came from, as the captain's widow still lives here (aged ninety-two), but could only glean that it was believed China was their home. Again in 1869 I was introduced to a Miss Hawker, an old lady, whose garden was filled with Silkies, bad, good, and indifferent. She told me that an eminent officer in the Navy (her brother) had brought them home for her, but again she did not know where from: she thought Japan. I obtained some half a dozen birds from her, and with them made a strain which has held its own until the present day. With much trouble a great fancier of this variety has been trying to find if the Silkies are found in Japan. He sent photographs of birds to see if they could be traced, but no one seemed to know the variety. I have quite recently had here, however, a gentleman who had spent twenty-seven years in Japan, and who told me that he had seen there fowls similar to those in my runs; and upon his sending photographs which I gave him to another friend who had lived some years there, the latter wrote: 'I have seen the fowls you mention in Yokohama and neighbourhood, but do not recollect them at Kobe or Nagasaki.' It seems therefore probable that the popular name of 'Japanese Silkies' is fairly justified.

"Some express difficulty in obtaining the beautiful ear-lobe and the deep-coloured mulberry comb in small birds. I have found the same. Nevertheless, it can be done, as my own experience knows. I grant that the coarser birds come with charming head-points and generally with excellent feet, while the feet and beaks are both blue, but the birds are too big. We have, therefore, to cross these birds with the smallest hens that we can find, and then, discarding all but the true in ear and comb, cross again once more these with the grand-parent or his brother. I have thus obtained a regular strain of beautifully headed birds. Some specimens have been winning during the past season with either lead-coloured ears or ears of no colour at all. I think that the points of the head should be most certainly taken into consideration as well as the silk.

The next question is that of green beaks and legs. The former are not so aggressive if they are only slightly tinged; yet they should be as blue as the beak of a cock Budgerigar. I have known a greenish beak in time grow blue; in fact, quite frequently so. But I have never known a green leg turn blue, not even a greenish one. Green legs show decidedly some *mésalliance* between strains. Green-legged birds I always recommend to be killed in their babyhood: it saves trouble, expense, and the dawn of hope which never comes to fruition."

SPANISH

History of the Spanish

The white-faced black Spanish has been much the longest known of the Mediterranean type breed, and it is perfectly easy to understand how it probably came to us direct from Spain. In the days of Philip the intercourse between this country and Spain was very great, so that Spanish and Portuguese wines almost drove French vintages for a time from the English market. It is further to be observed that at a later period, when Spanish were already known and bred in England, to a somewhat rough or cauliflower type of face, a second introduction of birds with smaller and smoother faces came from Holland, precisely that district of Europe which had been most over-run by the Spanish armies under the Duke of Alva. This crossing of strains considerably improved the English birds in face, as well as giving constitution; and two perceptibly different types of face remained till quite a late period, the late Mr Henry Lane, of Bristol, having bred chiefly the heavier kind, and only commencing a year or two before his death to transform it into a smoother character, which he saw to be more and more preferred by the judges. Another change which has taken place is in colour of the legs, which were many years ago desired as light as possible, and occasionally were brought nearly white, being put in poultices before exhibition in order to improve their colour! Yet another transformation must be recorded in the comb of the cock, which at one time was expected to fall over on one side, while now desired straight and upright.

The greatest change of all, unfortunately, is recent, and to be seen in a decline of popularity which has no parallel in any other breed. In *The Poultry Chronicle* of early exhibition days, there were more advertisers of Spanish than of any other variety; no fowl was so well known; none had so good a reputation as a prolific layer of large white eggs, especially in and round London. Now the breed has a class to itself at very few shows, is extremely delicate, and we fear it must be confessed, but a poor layer. The change has often been attributed to breeding too much for white face; but the facts do not warrant such a conclusion. The real reason has been lack of breeders, and in consequence a lack of blood, and an amount of in-breeding that has been ruinous.

Characteristics of Spanish

The Spanish cock should be tall on the leg, though not so stilty as most of the present day. The neck is long and gracefully arched, and the head carried high, with breast prominent; this proud carriage is apt to suffer from an overgrown comb, which is, however, less seen now in Spanish than in Minorcas. The shanks are slate colour or lead colour, but get lighter in old birds — almost a pale lavender in some cases. They are bred lighter now than some years ago, owing, we believe, to the more artificial regimen already alluded to. The body should narrow to the tail, somewhat like that of a Game cock. The tail should be full but not carried squirrel fashion. The plumage all over a rich black, as lustrous as possible. The chief points are, however, contained in the head and face. The beak is large and of a deep horn colour, and the head, as a whole, large, being both long, broad, and very deep in the side, with large eyes, which should be free and open in the midst of the face. The medium single comb should be perfectly upright, firm, and straight; rather thin at the edge, but thick at the base upon the broad skull; fine and smooth in surface, with a few broad serrations, not many narrow ones; and rising from the beak between the nostrils. But the chief feature of the bird is the white face. Both face and ear-lobe should be pure white, and in texture like the finest white kid glove; smooth, or free from ridges and folds, and leaving the eye unobstructed. The white should reach well on to the beak in front, rise over the eye close to the base of the comb, and extend well towards the back of the head, the further over and behind the ear the better, and sweeping in an unbroken curve towards the back of the neck. The large white ear-lobe should be long, open, and broad, lying spread out flat or free from folds, and not at all narrowing at the bottom, but keeping up the width till rounded off; thence the line comes up to join the wattles in front. These are long, thin, and florid, the inside of their upper parts and the skin of the throat between being white.

The hen is very similar in most points except that her comb falls entirely over one side of the face. The face itself is, of course, smaller than in the cock, but of the same general character; and there should be no apparent line or division between the face and the ear-lobe. The wattles are rather smaller than might be expected, and are preferred small and thin.

Breeding Spanish

In breeding for the main point of face, much judgement is required. It is better, as a rule, to mate smooth-faced cockerels, even if somewhat smaller in face (as such birds often are) with large and rougher-faced hens, than to employ the contrary plan, the produce of a male at all rough in face being very uncertain. Anything at all like a raw cross, even of good blood, often works apparent havoc in the faces; but nevertheless we would use a good bird in this way, in order to get, if possible, more constitution; as breeding back to "line" will make matters right in another generation or two. The greatest fault of the faces at present is not being flat and free from folds: so many are folded, wrinkled, or doubled. Something can be done to avoid this, by taking symptoms early, and treating by gentle extension at frequent intervals, but the root of the mischief is in selecting stock too broad in face on both sides. This tends to produce more face than the surface can really carry, and hence it folds or wrinkles up. Too thick or coarse a face is not only ugly in itself, but as the bird gets older very often grows so much as to obstruct the sight. In such cases a little has to be cut away with fine-pointed scissors.

It is not easy to determine the ultimate quality of the chickens while young, and we have known great mistakes made even by skilled breeders. The late Mr Jones will be remembered by many as one of the most successful: on one occasion he had ordered a cockerel for execution early in autumn. His "man", however, thought differently, and as the bird had a particularly handsome comb, kept him on for a bit to see what he would come to. He began to make up hand over hand, and turned out the champion bird of the year! Faces which show red, or even any blush of it, at an early age may of course be safely discarded. The best birds usually look a curious sort of blue in the face while young, steadily clearing to white as they grow older. The difficulty comes more in judging the ultimate size of face, which sometimes turns out much more and sometimes less than might be expected.

SULTANS

These birds belong to the great Polish family. Those now bred are from fowls sent to Miss E. Watts from Constantinople in 1854; but very similar birds appear to be mentioned in several old writers, and to

Figure 2.23 Spanish

be known in South Russia. Only one or two specimens have been imported since Miss Watt's originals, and breeders have never been numerous. Yet it is a pretty fowl, with very pretty ways and habits, and a disposition to accept and return petting; and a mixture of tameness and sprightliness, which is very attractive.

The crest of the Sultan is full and globular, the comb two tiny horns, beard and whiskers very visible. So far they are simply smallish white Polish fowls; but have five toes on each foot, the shanks being well feathered, and the thighs heavily hocked. The plumage is entirely white.

The following notes on the characteristics of Sultans are by the Rev. R.S.S. Woodgate, of Pembury Hall, Kent:-

"Although I have but of recent date taken to keeping this beautiful variety, still I have followed the breed for some thirty years with keen interest. It is a very charming fowl, with an immense deal of character. I have found them impatient if disturbed – in fact, they will fly out of their homes like birds, and alight on any adjacent tree or wall, only in two or three minutes to fly down again, and to be as domesticated as possible. They do well in confinement, and are layers of large white eggs, and these in goodly numbers. It would hardly be credited how large is the size of the egg from even a pullet of this variety, considering its size. The eggs also appear to be peculiarly fertile; anyhow, this had been my experience. I have found the Sultans to be most hardy, almost equalling the Silkie in this respect, while the chickens soon fledge and wander about over the dew coloured grass with pleasure and impunity.

"It appears to me a pity that this pleasing variety should not have more admirers, for the habits of the birds, whether in confinement or otherwise, are interesting in the extreme. I think, however, whether from in-breeding or otherwise, that the Sultan is hardly as massive or shapely as it was some twenty-five years ago. Anyhow, it will be a sad pity for the breed to be allowed to pass away, and a class or two at our big shows, at least, ought to show us what we still have in specimens of this beautiful and at the same time useful variety."

Figure 2.24 Sultans

The spurs of the cocks of this breed are especially apt to grow very long as the bird gets old, curling upwards so that the point enters the leg if left alone. Now and then we have seen a white Cochin cock in the same condition, but not so often as in the Sultans. When this occurs the spur should be partially sawn off and the point rounded.

Fowls resembling Sultans in being all white, crested, muffed and bearded, feather-legged and vulture-hocked, but differing in being distinctly high on the leg, were exhibited many years ago under the name of Ptarmigans, but have long been extinct. They were probably descendents of some former importation, the effect of long in-breeding in producing weediness of build being well known.

SUMATRAS

"The earliest mention of the breed that I have been able to find is in *Miner's Domestic Poultry Book*, published in America in the year 1853. They are described by Miner as of indomitable perseverance and courage, and noted for a beautiful green metallic lustre upon their plumage. He further gives notes by Dr. John C. Bennett, who stated that a trio of birds (and probably the first specimens) were imported to Boston direct from Angers Point, Sumatra, in April, 1847, by Mr. J.A.C. Butters. Dr. Bennett describes the birds then as having a small head, powerful beak, eyes lustrous, quick, and fiery, a pea-comb (though single combs sometimes appeared), small wattles with a very small dew-lap, hackles long and brilliant, tail long and drooping or horizontal (in the case of the cock with abundant 'plume'-feathers sweeping the ground; fan-shaped in the case of the hen), body slim and very symmetrical, legs sinewy, with a powerful and muscular thigh; colour of plumage variable, though he himself preferred black. The fowl was not known in England at this date.

As to breeding, Black Sumatras are not of large size, but it would appear that the birds do not now equal in size those of years ago. This, however, can no doubt soon be remedied by care and selection, but we must be careful not to obtain size at the expense of type. Both male and females in the breeding pen should, as far as possible, be perfect in this respect. The cock should have very long and flowing tail

Figure 2.25 Black Sumatras

and hackles, while if the tail feathers of the hens are very long and nicely curved, all the better. With regard to the head-points of most cocks there is room for great improvement, and care should be taken that not only does each of the hens have a strong beak, red eyes, neat comb and gipsey face, but that the cock has these points as far as possible also, though up to the present I have never seen a cock with the true gipsy face, and have been informed that this has not yet been obtained even in America. The colour of the plumage being a rich bottle green, as in Black Hamburghs.

THE SUSSEX

In the first impression of this work a hope was expressed that before it was too late some effort might be made to preserve from extinction the genuine *old* Surrey and Sussex fowl, which for years had furnished the very best table fowls to the London market, and which was really an ancient race, quite different from the various mongrels now prevalent in the district. We were glad to see that hope realised, and in July, 1903, steps were taken to form the "Sussex Poultry Club" for the breeding, exhibition, and standardising of this super-excellent old breed. By May, 1911, the Club included nearly 300 members. The birds were first exhibited under their proper name at the Royal Agricultural Society's Show in June, 1904, but, as might have been expected, were at that early date very irregular both in type and marking. But later, at the Lewes Show in November, 1904, the entries reached 163, and excited something like a mild sensation. The Hon. Secretary of the club is Mr. S.C. Sharpe, Brookside, Ringmere, Sussex.

That this old four-toed Sussex breed was one of the ancestors of the Coloured Dorking is absolutely beyond doubt. The late Mr. Harrison Weir took another view, but it is based upon statements and personal memories of men, each of whom thinks his own stock the only true one; whereas Bonington Moubray's evidence in 1815, and Nolan's in 1850, is contemporary and conclusive. It was of many colours, including speckles, browns, reds, and lighter plumage. From these the Club has selected for its standard colours a red, speckled, and light; the latter being marked like the Light Brahma. There is also a Brown Sussex Club, which supports that colour. The Whites, first bred by Mr. Godfrey Shaw as Albions, have been abandoned to the all-devouring Orpington interest; but it is a pleasure to observe that the speckled (which that interest also claims to appropriate) has been retained for the true type, and the county of its origin.

As regards type, the most distinctive characteristic of the Sussex fowl is the width and flatness of shoulders and back, in which it stands out from all other breeds. In shortness of leg, length and depth of body; and fulness of breast, it resembles the Dorking; but in this width and flatness of back it stands alone, and this feature makes the body appear almost short when viewed from above, though the breast is really long. This character is laid down in the Standard, and we are glad to observe that twenty points are alloted for the type, and twenty-five for size, as the breed is one for the table above all. The other main points are short white legs with only four toes, thin skin, and juiciness of flesh.

The Sussex fowl is hardy, both as a chick and later on, and the hen a very good layer. She is also a capital mother, and can shelter a very large brood, twenty or more chicks being often committed to one bird in its native county. But the commanding merit of this old British strain is as a table fowl, in which it surpasses every other breed on earth, even the Dorking itself. Both have white meat, and plenty of it; but the Sussex is by almost universal testimony superior in tenderness and juiciness whenever compared with the Dorking of to-day. Meall in 1854 is evidence that it was so at that date, when the race was plentiful; and it is noteworthy that since its remnants have been gathered together, Mr. Haffenden has won at Smithfield (December, 1904) first and cup in the class for any variety, with a pair of dead pullets in this breed. Many chickens will fatten in three weeks.

Mr. Lewis Wright's observations in our last edition we are now fortunately able to supplement by the follwing notes from Mr. S.C. Sharpe, Hon. Secretary of the Sussex Poultry Club:-

"Specialist clubs should have every possible support from all who are interested in the welfare of poultry, for the progress of the Sussex shows the good work that such bodies can accomplish, and yet only a few short years ago this most useful breed was scarcely known outside its own county, and not even there as an exhibition bird.

"The Sussex Poultry Club, with which I have been connected since its inception, now has two sister clubs affiliated with the British Club, one in Quedlinburg, Germany, Herr Heumann being the Hon. Secretary, and another club in America, Hackensack, New Jersey, Mr. W.H. Bratt being the Hon. Secretary. Both of these clubs are doing good work, and at the shows held in their respective countries some very good specimens may be seen.

"One of the three varieties of Sussex, the Light is my favourite — we have a standard for three varieties, viz. Light, Red, and Speckled — and there is no doubt whatever that these birds have been bred in the county of Sussex for a number of years before the Club was formed, and bred true to colour and type, too, and now that they are more widely known they have become a very popular fowl both in the show pen and as a useful utility bird.

"The Lights are capital winter layers; in fact for three consecutive years they have proved themselves best out of fourteen breeds which I keep at the Agricultural College Training Farm, Uckfield. As mothers and sitters they are hard to beat, bringing up large broods of chickens at all times of the year, and are used extensively in East Sussex for rearing the early spring chicken; and although good sitters, they are not so continually broody as some breeds I know.

Figure 2.26 Light Sussex Cockerel*

Figure 2.27 Light Sussex Pullet

WYANDOTTES

History of Wyandottes

As Sebright Cochins, or American Sebrights, a name which was also given to these birds, they never became widely popular, and it is somewhat uncertain whether this original stock did not die out. The name was not American or distinctive enough for popularity, and for several years it is difficult to trace any such fowls at all. But about 1880 similar large laced birds began to be freely written about again, and it is significant that regarding these were many actual statements, that the Light Brahma had been the Asiatic race employed. They were also termed Wyandottes, and the comb, while still rose in character, had assumed that downward curve at the back and of the spike, parallel with the top of the head, which is now a recognised Wyandotte point. This kind of comb would probably result from a Brahma cross; and upon the whole the evidence, though not conclusive, tends to show that some time before 1880 the Wyandotte had not only been re-named, but actually re-made, upon a Light Brahma and Hamburgh foundation, with possible aid from Polish stock, as "crested" birds are mentioned in one account we

***Modern birds would have very distinct neck markings.**

have seen. The first importation we were able to trace into England, was one by Mr J. Pilling, of Ashton, near Chester; and the first *English-bred* specimens to be exhibited were, we believe, shown by Mr T.C. Heath, at the Staffordshire show of 1884. The history of the breed in England, therefore, only dates from that time; and the progress made since, both in popularity and multiplication of varieties, is remarkable.

Quality and Characteristics

This is not without solid reason, for the Wyandotte is an undeniably valuable and generally useful fowl. It is a capital layer of tinted eggs, when bred with any reasonable care to maintain that property; is very hardy and easy to rear, feathering well and easily in chickenhood; is a capital sitter and mother, though not excessively broody; and is a very fairly good table-fowl. In this last respect it cannot stand so high in England as in America, where they prefer that yellow skin and shank which in England are rather disliked; but even in England it is beginning to be understood that a yellow-skinned bird may be excellent eating, and is sometimes more juicy than a white-skinned. It is a bird with capital breast and wings, at all events; and at the Smithfield show of table poultry in 1894, where all other breeds besides Dorkings and Surreys and Games had to compete in one class, the winning pullets were Wyandottes. In regard to laying qualities, it may be noted that in America, where fowls have been more persistently bred for laying than anywhere in the world, the White Wyandotte has slightly exceeded the average of any other breed, so far as we have been able to ascertain.

Apart from the plumage which distinguishes each of the varieties, the general characteristics of the Wyandotte are few, marked, and easily described. The head should be short and rather broad, the Brahma ancestry being here clearly traceable. The rose comb is smaller and narrower than a Hamburgh comb, and the back and spike or leader should curve rather downwards, parallel with the top of the head; this comb is typical of the breed, and should be preserved, otherwise it should be neat and full of "work" as usual in rose-combs. The face, ear-lobes, and wattles are smooth and fine, and brilliant red. The neck-hackle of the cock should be full and flowing, the back short, the saddle rising to the tail in a nice curve, the tail well filled up and sweeping, but rather upright. The body is very full and broad in breast, and deep, but not very long; rather what the Americans call "cobby" shape. This cobbiness, on medium or rather short shanks, is the characteristic type of the Wyandotte.

Silver Wyandottes

The recognised head and original of the Wyandotte family is the Silver or Silver Laced. In this variety, the head of the male is silvery white, the hackles lower down becoming striped with dense black, as also are the saddle hackles. The back is silvery white, as are the wing-bows; the principle wing-coverts white, broadly laced with black, forming good laced bars across the wing; the secondaries are black on the inner web, and white with a broad black lacing on the outer web, edging each visible feather with black. The breast and under parts are white laced with black, from throat round to back of thighs, the under-fluff slate or dark grey; the tail black with green reflections. The female also has a white head and striped hackle, and black tail, and secondaries of the wing as in the cock; the rest of the body white, with each feather laced round with black, the tail-coverts approaching that character as far as possible, or with a white centre to the feathers. Regularity and rich density in the black lacing is the main point in the value of the marking.

Breeding Laced Wyandottes

How to mate up to breed exhibition or standard birds is a question not yet solved in its entirety; but certain fixed principles guide all the better breeders in their choice of mates. The object of the fancier is to get chickens that will score the highest possible number of points in the show-pen. Experience has taught that whilst it is possible to obtain good cockerels and good pullets from the same pen, it is far easier and much surer to breed from two pens, one mated up to produce standard males, the other to produce standard females. This is the principle of the double-mating system. Much has been written against the double system, but nothing from the pen of any well-known successful Wyandotte breeder. In my own yards I once had a strain (Wood's) that bred both good pullets and well-laced cockerels from the same pen, but the cocks were inclined to be brassy-topped, although the pullets that were produced from the same mating never showed the least sign of soot or brassiness in their hackles. These brassy and sooty-topped cockerels I mated again to the whitest pullets, and similar results followed: namely, clear, well laced pullets, and faulty-topped cockerels. In fact, one of the best Silver pullets ever shown was bred from a cock very bad in top-colour. From this experience I deduced that top-colour in cocks does not affect the sound lacing in pullets bred from them, and in choosing my pullet-breeding cocks, I never regard the top colour now.

Another important item needs to be noted, namely, that pullets bred from a standard-breasted cock often run light at the throat. It will be found that the best breasted cockerels often have their own sisters with light breasts. An observant breeder, then, will not choose perfect-breasted hens to put in his cock-breeding pen, but those that are inclined to run light, and mate them to a perfect breast-laced cock on the dark side. It was this mating that produced my 1897 champion silver cockerel.

But the chief point to regard, or rather to disregard, in cock breeding, is the cushion of the hen. This may be mossy as a pepper-castor, and yet the bird be a splendid cock-breeder. In other words, the feathers on the back of the female have no influence on the progeny of cockerels in the saddle hackle; they influence top colour in no sense whatever. This should be apparent. The feather on the hen is laced on the back; the feather on the saddle of a cock is purely distinctive of the male, and finds no counterpart on the cushion of the female. Practice bears out the theory, and in breeding for cocks, mossiness in the hens mated up never troubles me, nor do I find the top-colour in the young cockerels bred from such is discoloured because of the faulty cushion. Cocks bad in top-colour are the outcome of sooty, brassy, or smoky neck hackles of the female, together with the already faulty saddle of the sire.

To breed good cockerels, I therefore choose (1) A standard cock of the heavy laced stamp. (2) He must be free from white in tail. (3) A hen with a well-laced breast on the light side, perfect wings, with a good black tail. But (4) do not regard mossiness on cushion; in fact, she is better for being dark near tail. (5) Comb, legs, shape, to be standard quality in each. Thus many a champion show cock has been bred from birds, none of which would gain a first prize in a second-rate show.

Pullet-breeding is a question I should prefer to have been treated by an abler pen than mine. That the art is not theoretically or surely discovered, is apparent from the fact that nearly every year the best pullets come from different yards, and from those, too, whose record for first-class birds has not been eminent. Breeding is a trifle flukey with most breeds, but that of Silver-laced pullets is uncommonly so. With a certain amount of diffidence, then, I simply offer the fruits of my own experience. It was just now remarked that standard-breasted cocks mated with standard-breasted hens are inclined to throw light-breasted pullets, and I wish now to emphasise that remark. In choosing a pullet-breeding cock I always select a dark-breasted cock, and that from a pullet-breeding strain. It is no use buying a cock for breeding pullets from a strain that has been used for years in producing exhibition cocks only, because the females in such a yard will most probably never have seen a show pen. Again, breeders always try to obtain the clean lacing of show pullets right into the tail. To produce such we must not regard too closely the exhibition points of the sire, and I would have no scruple in using a cock that had white in tail, and for choice would prefer one, especially if there were under the cock's saddle a number of clear laced pullet-feathers running into white in tail. But whilst unorthodox in these points, I like to be very particular as to fluff. A dark fluff is always correct, both for show and breeding purposes. A light fluff and peppery thighs largely account for double lacing, horse-shoe lacking, dullness in black, and other evils amongst the progeny.

There is only one thing to be considered in the female we breed from to produce pullets. That is, to get as perfect a show bird as you can.

Two practical hints may be added. Don't cross strains too much, unless you find that through continual in-breeding you are getting weak, and require more stamina in your stock. Then introduce new blood in the female line. And always have a few old birds in the breeding-pen. Young stuff breed bad feathers, lanky chickens, and narrow, long backs.

Golden Wyandottes

The Golden Wyandotte was at least several years later in date than the Silver, but all we are now able to ascertain is that Golds were written about as "new" in the American journals of 1885, and that in the winter of 1888 Mr Sid Conger came unmistakeably to the front with them, and that after this date they were largely bred by crossing his Gold cocks upon the Silver hens of other breeders. The first imported into England were sent to Mr A.W. Geffcken, and others soon followed, those early importations being considered fully equal, if not superior, to the then average of Silvers in Wyandotte type, and especially in comb. But there is not the least doubt that both comb and general type, subsequently to that, suffered considerably in this country from crossing with Indian Game, of which the signs were at one time very evident in many exhibits. Some used this cross for enriching the ground-colour, which became too dark in consequence; others, we believe, resorted to it with the idea of getting depth of colour in the lacing. At all events, we have seen many pullets which, in their sloping backs, narrowness at stern, and narrowness and hardness of feather, betrayed the cross most unmistakeably; and in pullets as late as 1899 we actually found *double lacing* — not that above described, of a light edge outside the black line, but the double *black* lacing seen in many Indian Game pullets. We knew one breeder, in fact, who bred and exhibited pullets from a cross of Indian Game upon Silver Wyandottes. The faults thus introduced, however, have now practically disappeared.

Buff Laced and Blue Laced

From the same fountain-head of the original Black-Laced Wyandottes, have proceeded yet two other varieties of singular beauty, known as Buff Laced and Blue Laced, or Violettes, which are closely allied, and in the main of the same parentage. The Buff Laced resembles in colour the variety formerly known as Chamois Polish, the ground-colour being some shade of rich buff, laced with white instead of black; the Blue Laced has a similar ground-colour, with a lacing of blue-dun or Andalusian colour.

Both these varieties originated in America, Mr Ira C. Keller appearing to have been first in the field. He commenced in 1886 to cross Golden Wyandottes with Whites, producing birds whose lacing was violet-blue; these violet-laced birds threw a certain number with white lacing, and from these were derived his Buff Laced breed, first shown at New York in 1895, and some of which were sent over to the Rev. John Crombleholme, in 1897. Mr Keller aimed at a golden or golden-buff ground-colour.

Quite another strain was originated by Mr Brackenbury, from entirely different materials. He mated a Golden cock with a hen of *solid* blue or Andalusian colour, produced from two generations of Golds on the male side, and of blue on the female side. She produced a pair of Golds with blue lacing, of which the female died, and the male was bred to Golden females, producing again blue-laced Golds. Meantime Mr Brackenbury had got from a cockerel bred from Golden Wyandotte and Buff Cochin, and pullets from White Wyandotte and Buff Cochins, a "sport" with Buff Laced plumage; and this bird was mated with some of the blue-laced females above mentioned. The cockerels from this mating all came black-laced; the pullets partly blue-laced and partly white-laced. One of the best *blue*-laced pullets was mated to a blue-laced male, and three-quarters of her chickens came creamy white with buff heads, and ultimately moulted out Buff Laced. The same bright-blue-laced pullet was afterwards bred to a Buff Laced (*i.e.* a white-laced) bird, and all of her pullets and three-fourths of the cockerels came white in the lacing, or of the Buff Laced variety.

White Wyandottes

Of the White Wyandotte, the last really pure or uncrossed descendent of the original Wyandotte race — for it was undoubtedly a sport from the Silver — the Rev. J. Crombleholme wrote as follows:-

"If I desired to keep Wyandottes for utility purposes only, I should select the White. The White, as a rule, is the plumpest Wyandotte grown, for, as there are no markings to breed for, but purity of white only, one need not fear to regularly introduce new blood in the yard. As a consequence, the enervation of constitution that follows too much in-breeding does not exist, and strong progeny is ensured. Another consequence of this freedom of choice is that the Whites are the best layers. Sweeping assertions of this nature are, perhaps, open to contradiction; at all events, my own best-shaped Wyandottes are the Whites; they are also my best layers, and produce the most fertile eggs.

"In breeding Whites we must insist on purity of colour. It is no use trying to get good chickens from sappy parents. There is something in a 'sappy' feather which no one that I know of has been able to diagnose, and which is always perpetuated in young stock. Anyone, then, anxious to breed exhibition chickens must insist on a true white colour in the parents. No matter how big or how fine a cock or hen looks, if they are yellowish or discoloured, keep them out of the breeding pen. When I first began

Figure 2.28 Wyandottes

breeding White Wyandottes I wrote to a noted breeder of White Leghorns, and asked him how he managed to show such extremely *white* birds, hinting that if there was anything in it he might let me know. His answer was, that his was a *white strain*. I took it then that he did not wish to tell me his secrets, and let the matter drop; but now, after eight years of breeding, I have come to the conclusion that this White breeder was not joking, but telling a straightforward tale."

The Black Wyandotte

Next to the White, among the self-coloured members of the Wyandotte family, the Black is the most popular. The reason is not far to seek, for the handsome contrast of red headpoints with rich orange beak and legs and bright beetle-green black plumage, makes a most attractive fowl. It is, moreover, well suited for its colour, docile disposition, and is a capital winter-layer to the poultry-keeper with limited accommodation and surroundings where the lighter plumage varieties would soon become dirty and unsightly.

As to type, the Black should be a true Wyandotte so as to differentiate it as far as possible from a Black Rock, and in other points allowing for colour the remarks on the White hold good. The yellow legs and beak, combined with sound black plumage, are the chief difficulties to be surmounted. Some breeders adopt double-mating; but others equally successful breed from the one pen only, and in the best interests of the variety it is to be hoped that this latter system will be generally followed.

Buff Wyandottes

Buff Wyandottes are avowedly cross-made birds, and were produced independently both in America and England. The first American birds were exhibited at Liverpool in 1893, but English strains were already in existence or being formed at that time. In America they were produced by crossing Silver Wyandottes both with Buff Cochins, and in some quarters with Rhode Island Reds, an American yellow-legged amalgam of Cochin and local stock, very similar to our own white-legged Lincolnshire Buff, now known as Buff Orpington. In England the Silver Wyandottes and Buff Cochin were chiefly employed. Owing to this further cross of the Cochin, Buff Wyandottes are rather apt to manifest more propensity to sitting than the other varieties; but in spite of this are remarkably good layers.

In regard to breeding the plumage, nothing need be added to what has been already stated in treating Buff Cochins, Rocks, and Orpingtons. As in the case of Rocks, the yellow shanks and beaks required, make the colour rather easier to breed than it is in Buff Orpingtons. The variety, as is natural, seems scarcely to hold its own in competition with Orpingtons and Buff Rocks.

Partridge Wyandottes

Partridge Wyandottes should have the exact colour and pencilling of the Partridge Cochin, with the shape and comb and legs of the Wyandotte family. They are of comparatively recent origin, but have become popular rapidly, and seem likely to remain so; the fact is that there is a natural fitness between certain breeds and certain colours, and the Partridge marking, or rather the colour which now passes by that name, appears to suit the neat and close-feathered Wyandotte type particularly well.

Many have an idea that the breeding of Partridge Wyandottes will be easy compared to Silvers and Golds. To such I say, 'Try'. I will give a little of my experience with the variety, which may help those that are taking them up to know how to breed, and what to breed for.

First we will take the cock. The beak ought to be a bright yellow, though very few are yet to be found that possess this point. Next, the eye should be red, or at least bright bay: a pearl eye is a great objection, and very hereditary. The comb should be of that true Wyandotte shape, termed a 'cradle comb', one that fits close to the head, and has the spike following the arch of the neck. Colour of head should be a rich orange, not red as we often see them. The hackle should be full, and fall well on to the back. The colour should be orange or golden-red, each feather having an intense black stripe down the centre, but not running up the full length of feather, or this will give the fault known as a smutty hackle, which is very objectionable. The breast should be a raven black right up to the throat; no feather should be tipped with red, neither should there be any red visible even when the feathers are separated. The black should also continue over the thighs, between his legs, and right up to the root of the tail; a bird that shows light or

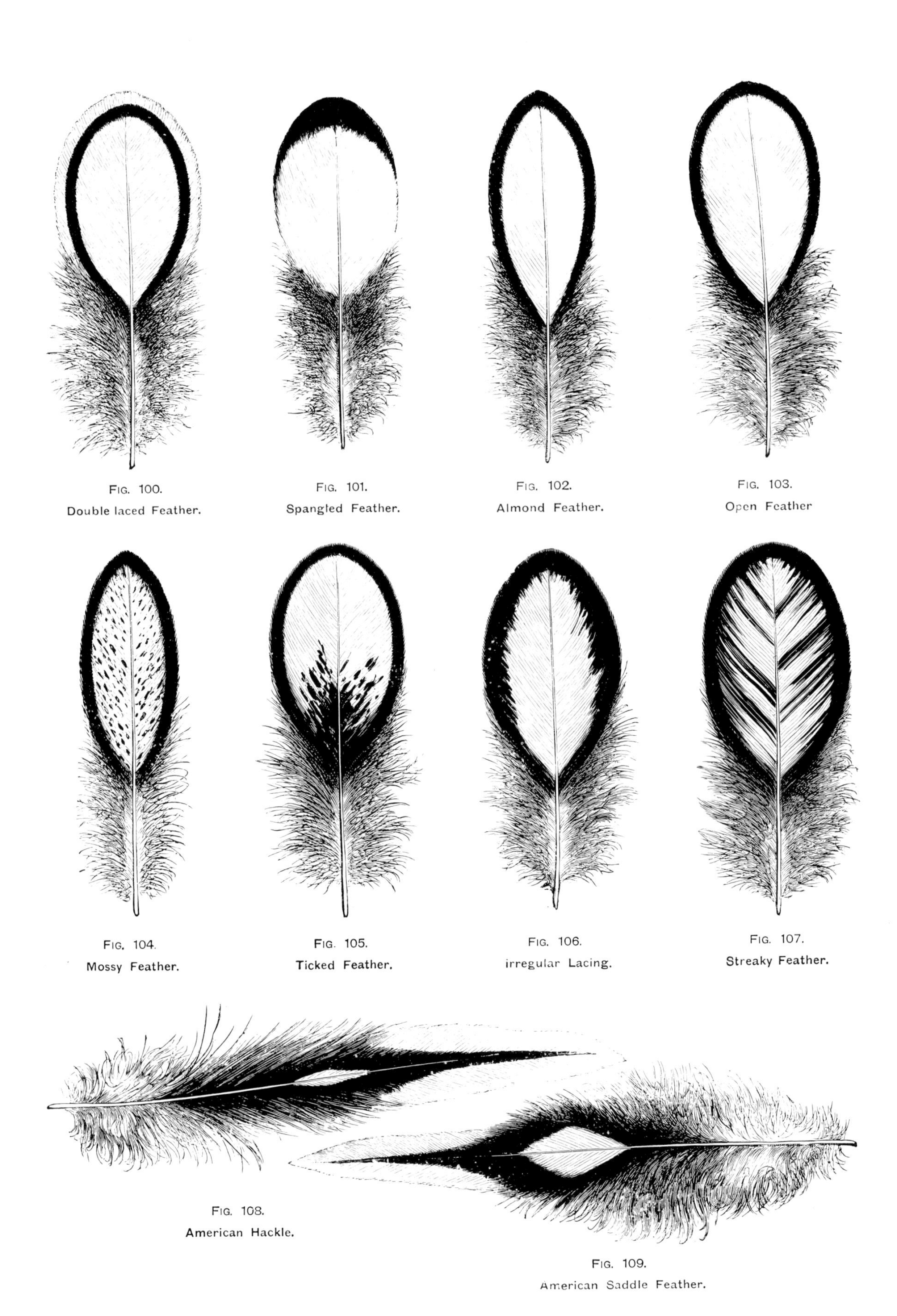

Figure 2.29 Wyandotte Feathers

grey behind is faulty. His back is required to be a rich red, but a bright red, not too dark a colour. The saddle hackle should harmonise with the neck, and be equally as well striped. The bar across the wing should be as black as the breast, free from red tipping. The secondaries should be a rich bay on the outer web, having a solid rich appearance when closed. The tail should be black right to the roots; many birds have white in tail. Legs, as in all Wyandottes, are required a rich yellow all round, not red up the side or smutty in front, with toes well spread and free from any sign of 'duck foot'.

To mate for the above we should require a male as near like it as possible. To him we would mate large hens or pullets, because without large females we cannot get large cockerels. See that these females have good combs, and decent coloured legs, with plenty of bone, and, moreover, a good Wyandotte shape. The neck-hackle is perhaps the most important point; see that these possess a good stripe, with an equally good orange edge. As we do not require pencilling in cocks, we should not look for it in his mates, and, indeed, we consider females without pencilling more likely to breed good exhibition cockerels. The ground-colour must, however, be of the desired brown shade. With the above mating we should get exhibition cockerels, but the pullets would only be suitable for again mating up to produce cockerels in their turn.

The pullet's beak, eye, comb, and leg colour should be exactly as in the male, but of course harmonising with her sex. The ground colour all over her body should be a rich light brown, not red, or it will be termed foxy; neither do we want it to err on the other side, or it will be termed grey or clayey. Each feather should be pencilled with a darker shade, the pencilling to follow the shape of the feather, as in the best Cochin hens. The pencilling should extend well up to the throat and right back to the tail, and with as much pencilling as possible on the thighs — in fact, pencilled all over except tail, which should be black, and hackle, which should be orange striped with black. Pullets with good yellow legs are few as yet, but breeders must strive for this important point.

To produce exhibition pullets we require a different mating from that of cockerels. The females must be as near the exhibition type as possible, and the male required is one that has been 'pullet-bred'. He will generally be one with tipping on breast and thighs, but here lies a great danger. You must know for a certainty that he is 'pullet bred', because faulty exhibition cockerels are sometimes sold as such, and such a male upsets all the mating.

One thing which makes these birds valuable from an exhibition point of view is the fact that their exhibition career is not over in a couple of months, like many breeds. The pullets moult year by year sharper and more distinct in pencilling, whilst the cockerels have generally a brighter top colour after their first moult.

Silver Pencilled and Columbian Wyandottes

Later varieties to be added to the family of Wyandottes are the Silver Pencilled and Columbians, and these bear the same relation to Dark and Light Brahmas that the Partridges do to Partridge Cochins. The former were originated by Mr George H. Brackenbury, the well-known American fancier and judge already mentioned in connection with Partridges, and were fashioned, as might be supposed, by mating the existing Partridge Wyandottes with selected Dark Brahmas, carefully selecting and "line-breeding" the produce. The first consignment received in England was two trios imported by Mr Wharton in January 1901. The three varieties are remarkable as presenting the beautiful Partridge and Brahma colours, with more moderate size, and freedom from leg-feather and fluff; thereby retaining the closer feather which the Brahma has now unfortunately lost, and with it the quality of heavy laying.

Both the name of Silver Pencilled, and contrast with the Partridge colour, have determined the choice of the whitest or paper-ground pencilling for these birds, and it must be bred in the same way as that type of Brahma. For details of this reference may be made to p. 20, and different pens will be required for breeding cockerels and pullets. To produce cockerels a bird must be chosen as good in show points as possible, and especially as clear as possible in white, as dense in the black, and with good striping in hackle. Want of clearness in white is the chief fault at the present stage. His hens must be of the same strain, and have solidly striped hackles and neat heads; their body colour may differ in various strains, but is generally rather dark: the blood is the main point. For pullet-breeding, on the other hand, we want hens or pullets with pencilling and show points as perfect as possible, and a cockerel of the same strain. He must have clear silvery and broadly striped hackles, and his breast will have either white ticks, or a laced edge at the tip of each feather, and on the fluff. If there be a narrow lacing of white in the tail or largest coverts, all the better. But again blood is the main thing, as a cock-breeding pen may also throw ticked or laced birds, which are, however, valueless for pullet-breeding. If the first year's breeding fails, breeding back to the sex the pen is selected for will usually succeed, if those birds are really good.

Columbians were first produced by breeding White Wyandottes to Barred Rocks, and the early birds were much wanting in striping of the hackles, and black in the inner webs of the wing feathers. This was gradually improved by care, and some breeders introduced Light Brahma blood, which of course remedied this fault a great deal, but brought occasional trouble with the leg-feather, and also too much fluff. The progress of the variety has been by no means rapid, and there is much yet to be accomplished before a perfect specimen is produced; but there is no doubt that better hens than cocks have been bred. The greatest improvement, within recent years, has been, perhaps, in hackle marking, both density of black and lacing being much better than formerly. Still, there are yet to be seen too many males showing canary-coloured tops, somewhat ugly combs, and fluff on legs. Hens, on the other hand, are of good type generally, and have much improved as regards hackle lacing, while body colour is good. With a little more density of black in hackle, and better lacing at the throat, the female would almost attain to ideal standard requirements; but this will take some time to accomplish.

YOKOHAMAS

About the year 1878 there appeared in Germany, and a year or two later in England, fowls imported from Japan, whose principal peculiarity consisted in length of tail and immense development of the cock's sickles and saddle feathers. Some were exhibited as Yokohamas, others as "Phoenix" fowls; but careful comparison of the representations published, and of photographs and drawings which reached us direct from the Continent, failed to show any distinction beyond greater or less development of their peculiar plumage. The long plumage was, however, unique, and a fair idea of it may be gathered from our illustrations. In a drawing of a pair of German birds in our last edition one of the cock's feet has five claws, a proof of the crossing which had undoubtedly taken place.

Correspondence brought out the fact that such birds had been occasionally exhibited as "Japanese Game" so far back as about 1872. It further appeared that in the Japanese Great National Museum at Tokio there were preserved two specimens of the race, in which the sickle-feathers measure 13½ feet and 17 feet respectively! And a feather has been actually sent to France which measured 2 metres 85 centimetres in length. In 1884, Mr Gerald Waller, of Twywell, imported a pen of these still more extraordinary birds; and from his statements it appears that they are known in Japan as Shinowaratao, Shirifuzi, or Sakawatao fowls, and by other various names. The very long-tailed ones are kept in high, narrow cages, always sitting on a perch covered with straw rope, with no room to turn or get down, but with a food and water tin at each end of the perch. Three times daily they are lifted down for a few minutes' exercise, their tails being carefully rolled up and enclosed in paper cases to keep them from injury. The Japanese state that a tail has been measured 23 feet in length, and that the birds only moult the tail once in three years. This last fact is highly interesting. It is obvious that if a tail 23 feet long were grown in one year, it must be at the rate of nearly three-quarters of an inch per day; and though Madame Bodinus states that she *could* "see the tails grow daily", it is difficult to realise this.

Tails exceeding 6 feet in length have, however, never reached Europe, the saddle-hackles of Mr Waller's birds having been about 16 inches; and in Japan itself tails over 6 feet are exceedingly rare. But the stock has had to be further crossed to prevent extinction. Nearly all that on the Continent was indubitably crossed with English Game, and from this the present known stock has been derived. By this somewhat has been lost in mere length of feather, but much has been gained in hardiness and real beauty.

Many names have been proposed. The Germans were mainly answerable for "Phoenix", which has no meaning; and "Japanese Long-tails" was too general. Some attempt was made to get Shinowaratao recognised, but to the appellation of "Yokohamas" the breed has now fairly settled down. We are indebted for the following notes upon it to Mr Frank E. Rice, of Sudbury, Suffolk, who has kept and bred it for some years:-

"To speak or write adequately on the beauty of Yokohamas (sometimes called Long-tailed Phoenix or Japanese Long-tails) is beyond the power of tongue or pen. They rank above all other breeds of poultry in their highly graceful character, and the beautiful formation of the tail, which is their special characteristic. The long sickle-feathers grow about 2 feet the first year, and each moult they come out longer, till the tails reach 5 feet and sometimes 6 feet in length. It is not altogether the length, but also the sprightly way in which they carry their tails: not in a drooping, dangling fashion, but in a most graceful curve from the formation which carries the weight, which adds perfect symmetry to an evenly balanced tail.

"The feathers should be broad and strong, except the hangers, which are soft and flowing, the saddle feathers hanging to the ground in great abundance of hackle feathers. Legs are medium length, of a bluish willow colour, and should have only four toes. In all respects the Yokohama should be a long-made bird, with long neck, long body, and long tail.

"There are several colours; those best known are the Duckwing colour and pure white, the former being the most attractive. At present all are exceedingly rare. The sickle feathers are used as plumes in officers' helmets and ladies' headgear, as in the former coloured birds they are of a most beautiful lustrous black.

"Notwithstanding their beauty, Yokohamas are very hardy and easy to rear. Chickens should be hatched in March, April, May, and June, to grow to maturity, as I have found very late hatched birds never grow much tail, which is such an important feature in the fowl. As layers it would be hard to beat them: wet or dry, snow or wind, they continue quite unconcerned. Their eggs are very rich, and although small, contain the same amount of nutriment as a full-sized one. The birds themselves are a dainty dish. Being exceedingly active, their eggs are wonderfully fertile. I very rarely have a clear egg, and find chickens hatch out very strong."

The birds that we have most admired have not been those with most length of sickle. Of this five to six feet can be got, as above described, but specimens of less age, and not exceeding three feet, have more impressed us.

Figure 3.1 Sebright Bantams

CHAPTER 3

BANTAM BREEDING

MODERN GAME BANTAMS

At no period in the history of the Poultry Fancy have Bantams reached such a popularity, or carried with them such a monetary value, as they do at the present time, so that show promoters in certain places are giving them exhibitions to themselves. Bantam breeders comprise, however, a fancy and world of their own, for which reason we have thought it better for this section to be treated throughout by some single special authority. Though not, as in other cases, distinguished by quotation marks, owing to their length, this and the following chapter upon Bantams are therefore written throughout by Mr P. Proud, of Birkdale, Southport. Mr Proud is so well known to every Bantam breeder, not only as an exhibitor, but as a popular judge, a critic, and writer upon this subject, that nothing further need be added beyond our gratification that he has found it possible to render us and our readers his valuable assistance.

The reasons for the great advance in popularity of Bantams are perhaps likely to increase rather than diminish. They can be kept as pets by ladies, young people, and others who would never trouble with larger breeds. They can be reared in hundreds, where medium-sized fowls can only be kept in dozens. They stand a town life well, and a few can be kept successfully in a small garden, or backyard, or, on a pinch, in pens in an attic. They are small consumers, and lay profitably for what little they eat. They are well catered for at all the best shows, and the returns from prizes won are, pro rata with the expenses of carriage incurred to and from exhibitions, greatly in excess of what could be expected from heavier breeds.

Hatching Bantam Eggs

In some respects Bantams require rather special management, and there are some special difficulties in hatching and rearing Bantam chicks. The selection of a suitable broody hen is highly important, unless the Fancier wishes to go in for a small reliable incubator, for such are now to be obtained from some of the best incubator makers; but an inferior machine is no possible use for hatching Bantam eggs, and the temperature for same must not exceed 102^0 For years I was greatly troubled by my hens breaking two or three eggs in every sitting, and crushing chicks when on the point of hatching, simply because they were too heavy and clumsy for Bantam eggs. At last I hit upon that cross between the Silkie and Pekin described in the next chapter. These pullets proved excellent sitters and mothers, and for hatching and rearing Bantam chicks are worth their weight in gold. They rarely lay more than a dozen eggs before becoming broody, and then they will sit till further orders; although it is not wise to let them hatch more than two broods at one sitting. As to the nests, where possible it is far the best to have a separate house for sitting hens, where the inmates can be left perfectly quiet. For nests I use orange boxes, which can generally be purchased for twopence or threepence. These I lay down on their sides, and nail a strip of wood along the front at the bottom. I then put in a plentiful supply of fine soil or sand, and on the top of this clean soft hay, the more the better, making the nest the shape of a shallow basin. In the early months — January, February, and March — never give the hen too many eggs; rather put down too few than one too many. Later on you may safely put down fourteen or fifteen. During the time the hen is sitting her food should consist of Indian corn, and she should be moved from the nest every day for ten or fifteen minutes. Before setting the hen give her a good dusting with insect powder, and again two or three days before she is due to hatch; by so doing you will save many a chick from being infested with lice. In the sitting-house have a large shallow box filled with cinder ashes for the hens to dust in.

Figure 3.2 Various Bantams

After five or six days the eggs should be tested by candle in a dark room. Hold the egg between the first finger and thumb of each hand about an inch from a lighted candle. If the egg is unfertile it will appear quite clear, whilst if it contains the germ this will be found to float and show the veins quite plainly. At the end of nineteen days, if the eggs are fresh, they will commence to hatch, and at this time the hen should be left alone; do not bother her more than you can avoid. When possible always use fresh eggs for hatching; and never more than eight or ten days old, if you would have strong chicks. Never set very small eggs, or thin-shelled eggs, for even should they hatch the chicks will be fragile little mites that it is impossible to rear.

Game Bantams

No one will dispute the fact that of all Bantams the Game varieties must take precedence. Of all the different kinds bred, none seems to secure the popular vote like these. Fanciers may drift into other varieties, but as a rule they begin with Game. Fearless little creatures they are, with a pluck and courage all their own. No wonder they stand first favourites! The colours allow a liberal margin for the exercise of individual taste, though Black-reds, Duckwings, Brown-reds and Piles have so far made the greatest advance towards perfection, followed by Birchens and Whites. As a rule Black-reds and Piles give the heaviest returns for the capital and labour expended upon them.

In judging Game Bantams, shape and style come first. In both male and female we want somewhat of the laundry flat-iron type, i.e. a wedge shape, with no great keel, and tapering from a square prominent front sharply to the tail. The shoulders should stand out squarely, the wings short and rounded and tucked well in to the sides, not flat-sided. If the wings are carried on the back, the bird is said to be goose-winged, which is a serious fault in the show-pen. A round back is an eyesore, and practically a disqualification. The thighs must be round as well as long. A flat-shinned bird comes of delicate parentage, and will breed flat-shinned ones. The thighs should be set on well apart, as this affects greatly the Game-like look of the movements of the bird. The shanks must be long, fine, and round in bone, clear and smooth, and terminated in long straight toes with good claws. The hind toe should be planted firmly on the ground in a straight line with the centre front toe, any screwing about to right or left, or inwards, is termed duck-footed, and is a certain disqualifiction in the show-pen. Sometimes in very reachy birds the back toe proves a trifle short, but this is a minor fault compared with the other. The tail is a special feature. It should rise to a little above the horizontal, and the twelve feathers of which the tail proper is composed should be of the shortest, narrowest, and scantiest; and in the body and tail feathers as strong in quill and as wiry to the touch as possible. The sickles and hangers have been bred to a surprising narrowness, and very short, so that the former only protrude beyond the tail proper about a couple of inches. The head should be long, lean, and snaky, with a long, strong, slightly curved beak, and a full, defiant, eagle-like eye, placed just under the top of the skull. Look for a long and fine neck, with bright, narrow, short hackles, and very upright or perpendicular carriage.

Colour is a great desideratum, and should come next. A rich colour, giving wide contrast, greatly enhances the appearance of a bird. In the males there should be no rustiness, especially about the shoulders, wing-bars, or hocks. The standard colours of the different varieties will be described as we proceed.

Size, or want of it, is a very important point, but diminutiveness may be overdone. All other points being equal, the smallest bird would undoubtedly win in the show-pen, and for this reason small birds are much in request, and command excellent prices. But very small pullets are by no means desirable in the breeding-pen. Three possibilities may arise. The first is they may never lay at all; the second is that their first egg may prove their death; and, lastly, should they produce a few eggs, the chicks resulting will in all probability turn out such stunted little specimens as to be worthless. Breed from small cocks and fair-sized hens or pullets, feed judiciously, and you will secure chicks small enough for keenest competition.

Black-red Bantams

The colour of Black-reds may thus be described. The face, head, lobes, and wattles in both sexes should be of a bright healthy red, with legs, feet, and beak of a rich olive or willow. Any tendency to slatiness shows Duckwing blood. The neck-hackle of the cock should be a bright golden orange, free from striping, and the saddle-hackle follows suit, only as a rule in the best specimens the saddle-hackle runs a trifle lighter, clearer, and richer golden hue than the neck. The back and wing-bow are a bright solid crimson; wing-butts black; wing-bars a glossy steel blue, free from rusty ticking; the secondaries of the wings a clear bright chestnut, which should run through to the end of the tail a rich black, free from ticking, rustiness about hocks, or chestnut shaftiness in sickles, which generally denotes a pullet strain, and will tell slightly against an exhibition specimen.

The Black-red pullet should have a pale golden-coloured hackle, some shades lighter than that of the cock, each feather being very narrowly striped with black down either side of the shaft. The main body colour is somewhat difficult to describe. It resembles the medium brown drab shade of a partridge. If each feather be examined carefully, it will be found to be almost imperceptibly pencilled through with the finest black, yet so exquisitely fine as to produce one even soft shade of colour all over back and wings. Imperfect specimens often get this pencilling too much in evidence. Such are termed "coarse". Even in some of the best specimens there are two or three feathers, usually the top flight feathers of the wing, that show the undesirable blotching a little round the edge of the feather; but this is considered rather a serious fault in exhibition pullets. The breast may be described as a rich deep broken salmon, shading into lighter colour towards throat and thighs. The two top outer feathers of the tail correspond exactly with the colour of the body, otherwise the tail is black.

Pile Bantams

Piles are treated of next because of their striking appearance, though amateurs will find them more difficult to breed than Black-reds or Duckwings. The Piles are also the counterpart of Black-reds, if we substitute white in the cockerels for black in the Black-reds, and a clear white in pullets for the partridge-brown of the Black-red pullets. They can for a certain length of time be bred solely from Piles; but it is found in practice that the colour "breeds out" after some five or six seasons, and it is necessary to resort to Black-red blood after a time. The exhibition standard is met in cockerels by a top-colour identical with that of the best coloured exhibition Black-red cockerels. Some judges prefer a deep claret back and saddle shading down to a golden orange; but I much prefer the former. The breast and bays are two very important features in the cockerels. There should be no lacing, or ticking, or smokiness in the former; whilst the bays should run right through the secondaries to the end, leaving no white or paleness, but one deep, rich chestnut. The eye should be kept as bright red as possible — a more difficult task than in Black-reds. The pullets require to be free from foxiness on the wing, but at the same time, the breast should be a rich salmon, and to get the two points in conjunction tries the utmost resources of the breeder. With the deep breast, as a rule, you get more or less rosiness or foxiness on wings, and with a paler breast you are more likely to secure clear colour on wing. Still, in these days of keen competition, a high-class bird, to be successful at our best shows, must excel in both points, and they can be obtained by patience and careful mating of pure pullet strains. Legs other than a rich orange are a most serious blemish. Years ago willow-legged Piles were successful in the prize lists, but those days are over, however good a bird may be in all other points. A rich deep orange leg is somewhat difficult to obtain, especially in certain parts of the country. After repeated experiments, I have fully satisfied myself that the maintenance of rich leg colour, when once obtained in a chick, is dependent in no small measure on the nature of the soil. Clay, good loam, and sand are helpful; limestone seems to fade and bleach the colouring matter considerably. A good, rich leg colour is indicative of a recent cross of Black-red blood, and when leg colouring begins to fail in any young chicks it is a sure proof that a cross is required.

Duckwing Bantams

Duckwings next demand attention. They are lovely coloured birds, more easily bred than Piles, and command fair sales at good prices. Up to the present only the Golden variety is tolerated in the show-pen in this country. The male is easily described. In every respect save top-colour and wing secondaries, which are white, he is coloured as a Black-red. He must have the same cherry-red face and appendages, including the most important factor — a red ear (not one trimmed up with scissors). In shape, style, reach, eye, and quality of feathering, he must be identical with the Black-red. Then for bright crimson on shoulders and back, substitute a deep straw colour, more or less shaded with maroon, which gives a very bright brassy look right across the back of the bird down to the tail, shading off to a creamy white on saddle, matched by the same colour on neck. The hackle feather both on neck and saddle should be as free from dark striping as possible. The secondaries of wings should be a clear white running right through to end of feather, and free from chocolate marking on the outer edge of the top feathers. A great many otherwise good birds fail here.

The pullet for exhibition is distinguished from the Black-red by the substitution of a lovely pale slate or steel-grey colour; this is exquisitely and finely pencilled over in the best specimens, but coarseness will show itself in indifferently bred birds, and the fatal blotchiness will in such also appear on wing-ends. Some strains will show a tendency to shaftiness of feather, which is also a defect. The back and wings should show a delightful softness and eveness of marking all over. The legs, as in the cock, should be willow. One of the most difficult points to secure in her is the deep salmon breast, with a soft, even body-colour of the lighter shade. With a light body-colour the breast is apt to run pale. If very pale, then it militates greatly against a show bird, but a fairly deep breast is no serious drawback.

Brown-red Bantams

The **Brown-red** is a very taking variety, but is not making much progress, and has never found so much favour as the three preceding. These birds are in too few hands, and suffer considerably by reason of their want of distribution. Pullets seem in advance of cockerels, and I have seen some lately which, for both type and style, could creditably show the best Black-red pullets the way. The chief fault at present is too much feather and a certain softness of feather, which seems incidental to the pale lemon colour. The little Brown-red lends itself very accommodatingly to the requirements of those who live in smoky and dusty places, and can be kept where a Pile would be smoked out. They are also easier to breed than either Piles or Duckwings, and with a little management could easily be bred from one pen.

Both sexes should have dark, mulberry-coloured faces, "gipsy" as it is termed, and the eye should be as dark as it is possible to get it. A light eye is indicative of a cross, as is also a red face, both serious faults in the show-pen. The legs and feet should be almost black. The neck and saddle hackle of the cock should be a light lemon colour, not orange, but more of a mustard colour, with the back and saddle a richer, deeper lemon; breast a black ground-colour, with a pale lemon lacing (sharp and well defined) round each feather, and extending well down to the thighs; wing-bars and secondaries black, free from lacing, tail green black. As a rule the feathering is neither so fine nor so short as in Black-reds, these points having been sacrificed to obtain the light top-colouring, which a few years ago was more of an orange tint than a deep lemon. Such a bird now would have no chance, even if harder in feather, so much does the lemon craze rule everything. In type, shape, and size, the Black-red ideal should be sought, though as yet some of the best Brown-reds are rather too large.

The exhibition hen should have a pale lemon neck hackle, with a very narrow stripe of black running down each side of the shaft. A most important item is that she should not be coppery capped. It is one of the great difficulties that engage the breeder's attention to secure the pale lemon lacing from crown of head downwards, as such birds are apt to be also laced on shoulders and back. The breast, as in the cock, should be exquisitely laced with pale lemon from the throat right down to the top of her thighs. The body, wing, and tail should be a glossy black; style, shape, and size as in Black-reds.

Birchen Brahmas

Birchens today are more popular than they ever were, but still not more so than I should like to see, for really there is no prettier or more taking bird than a good Birchen pullet. As already stated, they came orginally by crossing Duckwings with Brown-reds, but now it would seem preferable to breed them as much as possible *inter se,* so as to get the proper mulberry face and deeply coloured eye emphasised as much as possible. The male bird is easily described, as he resembles the Brown-red in every point save top-colour and lacing on breast, which should in both cases be a pure silvery white. As there is Duckwing blood, the greatest care should be exercised to eliminate the tendency to red face and eye resulting from a cross with a "red-faced" variety. The pullet also resembles the Brown-red in every point, if we substitute white lacing of the breast and neck hackle for the pale lemon of the Brown-red. Naturally there will be some tendency to lemon hue in the neck. This should be rigorously suppressed, and is more in evidence in cross-bred than in pure-bred birds. Another serious fault with many pullets is a dark cap instead of a uniform silver white from crown to end of hackle; and even some of the most typical birds hitherto have failed in sparsity of lacing on breast. The difficulty is, when the breeder has secured the lacing from throat to top of thighs, to prevent it appearing elsewhere, as on back and wing, with shafty feathering, which would condemn a bird much more than the want of enough lacing on breast, though such birds would prove gems in the cockerel-breeding pen. Of course the eye will give trouble; it is only to be expected. Red eyes should be rigorously eliminated; though we often see birds with this defect winning, it ought not to be so. A deep brown or black eye is the correct thing, though it seems next to impossible to get them as coal black as in Brown-reds, owing to the Duckwing cross in them.

The mating up presents no difficulty. The same lines should be followed as with Brown-reds. It is preferable to breed from pure Birchens, but if unable to procure these, one must begin at the beginning, and procure two or three typical Brown-red hens or pullets heavily laced on breast, and mate them to a Silver Duckwing cockerel of good quality.

OLD ENGLISH GAME BANTAMS

After lying for years in a dormant state the fancy for Old English Game Bantams suddenly leaped at a bound into amazing activity. The breed was as old as the hills, but had been much overlooked since poultry shows came into fashion. I remember having a fine pair when I was a boy of some ten summers, over forty years ago. They were of the "spangled", or, as it was then termed, the "speckled" variety, and handsome birds they were, though perhaps a little larger than those which would nowadays grace a show-pen. But they were by no means new then, for my grandfather kept them in his day, and the probability is that his forefathers some generations back had them too. Latterly they have come with a rush again, and no committee need fear that their classes will not be filled with this breed. And as to prices, what might have been purchased a little while ago for a shilling, and picked up in a backyard, compasses now a couple of £5 notes.

The varieties of Old English Bantams, like those of Old English Game, are legion, but in my estimation the palm for beauty must go to the Spangles, followed by the Black-reds. In no way are the dwarfs inferior in pluck and defiant attitude. I feel sure that they will live long and see good days, which they well deserve. They are extremely hardy and healthy, easy to rear as chicks, and may be kept in exposed situations where the Modern Game Bantam would perish, whilst today, in 1911, they are certainly more popular than Modern Game Bantams.

I will now try to describe what in my opinion an ideal Old English Game Bantam should be. Head medium length, but thicker than the Modern Game Bantam; face bright red, with a red, fiery, defiant eye, strong, slightly curved beak, comb small, of fine texture and single, wattles and ear-lobes a bright healthy red, the latter without a trace of white. Neck-hackle plentiful, coming down well on to the shoulders, and covering a fairly long neck, set well between a pair of broad, prominent shoulders, and above a full, broad, deep chest. Back broad and short, stern fine, giving the body a wedge-like look, wings short, well tucked into the side, but full, so as to avoid any appearance of being "flat-sided", which is a serious defect. Legs short and thighs muscular, well set apart, shanks medium length and round in front, not flat. The legs should be white or yellow in the before-mentioned Spangled variety, but white for preference, toes long, straight, and muscular, with no signs of being duck-footed. Carriage bold, sprightly, defiant and independent. The tail should be the complete antithesis of that of a Modern Game. The square, or hen tail as it is sometimes termed, is longer and broader in webbing, whilst the sickles are broad, a good length and nicely bowed, with four or five good side hangers on either side placed so as to well clothe the tail proper. The bird when in hand should feel corky and yet hard.

The points of colour in Black-reds are exactly those of a good Modern Black-red, with beak to match the legs. The tail, and also flight feathers, sometimes runs into white, which is to some extent a defect, but only a slight one, in the exhibition-pen. The pullet is either of the partridge or wheaten type, broad in front, short in back, with short, muscular legs and a full tail. The partridge colour has been fully described already, under the Modern Black-red Bantam hens. The Wheaten is a beautiful bird, with a bright golden hackle, with narrow dark striping down each side of the shaft. Her breast and thighs are a pale fawn colour, whilst the top colour, together with the two top outer feathers of her tail, is that of red wheat, hence the name. The tail proper is black.

The colour in Spangles is very beautiful, both sexes in this respect being identical. Heads cherry red, as in Black-reds, plumage throughout black, red, or blue, evenly spangled all over with white, tail black and white, legs white or yellow.

The cocks run from 14 ozs to 20 ozs, whilst hens are from 12 ozs to 18 ozs.

Breeding Old English Game Bantams

Black-red cocks may be Wheaten-bred or pure. If from Wheatens, the colour is not so bright, strange to say, as from partridge, if the cock be partridge-bred too. The brightest golden-coloured cocks, partridge-bred, are much brighter than the brightest from Wheatens, and are generally a sounder black on breast. The partridge hens must, of course, be bred from pure partridge-bred birds on both sides, with no admixture of Wheaten blood whatever, and the same lines should be followed exactly as recommended for the production of the Modern Black-red. To successfully breed Wheatens we want Wheaten blood on both sides, a Black-red Wheaten-bred cock with Wheaten hens or pullets, and this cock should be much darker in top colour than the exhibition cock, and be pure Wheaten-bred.

The best Spangles seem to be produced from parents evenly spangled on both sides. Should the produce, however, run too light, use a partridge hen, as producing a rather more pleasing shade than a Wheaten, and harder quality of feather. My choice, however, would be to breed from evenly spangled birds rather than to resort to this cross.

Blue Duns require a word or two, though no very precise lines can be taken for their production, as they may come from Black-red cocks and blue hens, or a blue-red cock and a Wheaten hen, all being very sound in colour to begin with. Probably a good blue hen could be bred from the latter, as well as from pure blues on both sides. The fact is they can be bred many ways, and it hardly matters how, as, if proper type, shape, and size be secured, colour in an Old English Game fowl of any kind is quite a secondary consideration, with the exception of Black-reds. What must not be, is any attempt to foist upon the judge a thick heavy waster of the Modern type, as an Old English Game Bantam. It has been tried, and in some cases with success, but new century requirements are much ahead of the last couple of years of the last century, and as there are now hundreds of the genuine thing in the land, there is really no necessity to try such deception, which would be now instantly detected by a good judge.

THE "VARIETY" BANTAMS

Black Rosecomb Bantams

Black Rosecombs are fairly hardy if not too greatly in-bred. The chicks, however, often require special care for the first few weeks of their lives, after which they forge ahead, and are able to stand a decent share of knocking about at shows, and indifferent weather. They are capital layers, but a cold, heavy, clayey soil is against them. A dry, sandy, and somewhat shady locality is the best.

We want a rose comb fitting close to the head, fine in points, full of work, bright coral in colour, and finished off with a long, fine, round leader or spike behind. See that the comb is square and full up in front, with no valley down the centre, or leafy in front, both bad defects. The face must be a bright cherry red, entirely free from white, which often shows its approach by a tiny white speck under the eye or round and under the white deaf-ear. This deaf-ear or lobe must be round, thick, large, and white, of smooth, kid-glove-like texture. The wattles are bright red and well rounded; eyes dark and full; head short and fairly broad, with dark beak, slightly curved; neck short and thick, and heavy in feather, spreading well over shoulders; back short and broad; wings not too long, but fairly large, and carried rather low; tail as full of broad feather as possible, both in the tail proper and in the hangers and sickles, which cannot well be too long or too broad, or too many in number. The bird throughout requires the most ample flow of feather, and the carriage is jaunty and important. Legs and toes black, short, and fine, with either white or black nails. Breast broad and prominent, carried somewhat Fantail fashion, with head well thrown back. In adult and old birds the legs will often be found slate or pale coloured. The total weight should be from 14 ozs to 18 ozs, and in Blacks the green sheen should be one of the characteristic beauties of the bird.

The hen should have the same red face, comb, and wattles as the cock, only the two latter should be much smaller than in the male. See that the comb is not flabby, loose, and coarse. The ear-lobes must be rather large, with beautiful texture. The breast is full, broad, and prominent, as in the cock. The back should be extra short; tail full, and carried rather gaily; whilst legs are short and dark. The whole carriage of the hen is strutty and important. The wings may be fairly long and drooping as in the cock, and as much sheen on her raven feathers as possible should be secured.

Sebright Bantams

In the Sebright or Laced Bantam we have another most beautiful variety. If the ladies cannot take to the Sebrights I shall lose all faith in them (the ladies I mean, not the Sebrights). The birds are prime favourites at shows, and invariably attract a lot of attention, both from their style and carriage, and also their beautiful lacing, which in good specimens is almost perfect. This breed is no new manufacture, dating from the days of Sir John Sebright, who worked with it indefatigably, and got together a host of enthusiastic friends who took it up, and had special delight in competing with one another in their frequent exhibitions. It matters little how it was derived, but it probably was of Polish descent, differing from the larger laced fowl in absence of crest, and of sickles in tail.

There are two varieties, Silvers and Golds, but at the present time Silvers are by far the more popular, probably from the fact that the contrast is greater in them, and in consequence more catching to the eye, and also that they are much easier to breed true to colour. In both varieties it is not so much the lacing that is difficult to get, as the quality of it. We often see it a dusky, rusty colour, instead of a good beetle black, and especially so in the Golds. It requires much care in this case to get a good rich ground colour, combined with the necessary quality of glossy black, at the same time maintaining an absolutely clear as well as rich ground. Often the Golds run far too pale in ground-colour, which again is a serious fault. This can, however, to a certain extent be remedied by colour feeding.

CHAPTER 4

DUCKS AND ORNAMENTAL WATERFOWL

DUCKS

The domestic varieties of ducks are much more numerous than in 1861, when Mrs Fergusson Blair wrote in *The Henwife* that "there is not a great variety in our domestic ducks; only three distinct exhibition breeds exist, viz. the Aylesbury, Rouen, and Buenos Ayres or East Indian". Though, curiously enough, the last named has now practically disappeared, being replaced by the Cayuga, the list now would be twice as long; but the first place, at least in England, must still be given to the Aylesbury breed named from the county town of Bucks, which has for generations been the chief, and is still the largest centre of the duck-rearing industry, and where scarcely any other kind of duck is thought of in connection with it.

Aylesbury Ducks

The Aylesbury duck is long in the body, which is carried horizontally, the legs appearing almost in the middle. The neck is fine and rather long, with a somewhat swan-like carriage, the head a little snaky, with the bill long and coming out straight like a woodcock's. The most obvious characteristics to catch the eye are the pure and spotless white of the plumage, entirely free from the least yellow, which may contaminate Pekin-crossed birds, and the delicate colour of the bill, which should be a pale soft pink, "like that of a lady's finger-nail". This colour is partly the result of long and careful breeding; but there appears little doubt that it is partly due also to constant scouring in a fine shelly gravel found throughout the Vale of Aylesbury about the streams and ponds, of which the ducks are very fond; and it is only found in perfection when the birds are kept out of much sun, and away from ferruginous soil or foul water, which injures the colour greatly. Left to wander at pleasure in other localities, the progeny of the best stock will often turn yellow in the bill, but it is found that this can be prevented by care, and by placing in their troughs abundance of fine white sharp gravel. The legs are bright orange. The drake differs in no respect from the duck, except in being rather larger, and having two or three curled feathers in his tail; and at an early age it is difficult to distinguish the sexes, as these male feathers do not develop till the first moult. As the young ones grow to the age of six or eight weeks, however, the voice of the two sexes will be found to differ, the ducks giving a distinct "quack", whilst the note of the drake is not a quack at all, but much fainter, and husky in character.

In one respect the Aylesbury duck has somewhat changed during recent years. As we remember it in 1860, and for some years afterwards, it was not what duck-breeders term "keeled" underneath the body. By degrees this character was more and more cultivated, from a desire to increase the massive look of the birds, and at the present day the Standard describes and requires keel — the term is self-explanatory — in exhibition birds. To a certain moderate extent, while confined to a fairly deep breast bone, the point is useful; but the present exaggerated keel is disliked by market dealers, and has had something to do with a partial displacement of Aylesbury ducks by Pekins in the London market. The average weight of good stock is about 7 lbs for drakes and 6 lbs for ducks, in ordinary condition at twelve months old; and such weights, if of large frame, are quite large enough to breed from. The heavy weights of 10 lbs and 9 lbs seen at exhibitions are obtained by forcing diet; and birds once fed and fattened up to it are practically worthless as breeding stock afterwards. As if that were not enough, both these and other ducks are often *crammed* before judging, or tempted to eat a pound or more of raw sausages, or meat; or some exhibitors even provide a quantity of live worms! Some ducks have been killed in this way, but a moderate quantity of uncooked sausage, or something of the same kind, often seems to pick up the birds and make them look better in every way after the journey.

Rouen Ducks

The Rouen duck closely resembles the wild Mallard in its plumage, except that this has been bred richer and darker. It is very probably correctly named from the city of Rouen, as ducks more or less resembling it in colour are still plentiful in Normandy, though not bred to such a precise standard of feather, or so massive. In general conformation the Rouen is somewhat shorter and deeper in body than the Aylesbury, and considerably deeper in keel. The bill should be long and broad and straight, as in that breed, that of the duck, however, being rather shorter than the drake's. The drake's bill should be a greenish yellow with a black bean at the tip, lead-colour amounting to disqualification, and too bright a yellow being also disliked. The head is a rich green, glossed with purple, which extends down the neck to a collar of pure white; this does not quite meet at the back, but must be clear and distinct so far as it goes. The breast is a rich deep claret extending down well below the water-line, and free from the fine white lacing which is called by breeders "chain armour". There it joins the delicate French grey of the flanks and under parts, which should extend to under the tail, any pure white under the tail being a great objection. This French grey is minutely pencilled all over with fine black lines. The back is a rich greenish black, the curls in the tail being a dark green. The wings are greyish brown, with a "ribbon-mark" across them, which must be a very bright and distinct blue, edged on both sides with black and white bands. The flights are grey and brown, white in a flight-feather being highly objectionable. The drake's legs are a rich orange.

The bill of the duck is of an orange colour, with a splash of nearly black upon it, two-thirds down from the head, but not reaching the base, tip, or sides; this colour, however, changes during the laying season to a dirty brown, and sometimes they become almost black all over. The head is brown, with two distinct shaded lines on each side, running from the eye down to the darker part of the neck. The breast is brown, pencilled over with dark brown; the back pencilled with very dark brown, or black glossed with green upon a brown ground. This pencilling must be very distinct, though judges differ somewhat as to the shade of brown which should form the ground-work. The wing has a ribbon-mark, as in the drake, and the legs are like his, orange, but of a dusky shade.

Pekin Ducks

The Pekin duck is a comparatively recent introduction, and one of the most valuable. It is rather doubtful whether the first importations were made into England or America, Mr Palmer in the United States and Mr Keele in England having both imported birds in 1873, and both exhibiting them in 1874; but in each case there is no doubt they came from Pekin, and were from the first a most well-marked variety, though they bred freely with other ducks.

The Pekin duck differs from others in the shape and carriage of its body, in a full spherical growth of feathers under the rump, and a singular turned-up carriage of the tail, the whole irrresistibly suggesting the outline of an Indian canoe. The legs are set far back, which makes the bird walk rather upright or penguin fashion. The neck is somewhat long, and the head decidedly large in proportion. The legs and bill are a rich yellow or reddish orange, the bill being shorter than in the two preceding varieties; and the plumage approaches white, with a peculiar canary yellow running through it. At one time some exhibitors showed pure white, but such specimens nearly always had pale bills, and there is no doubt that they originated in a cross with the Aylesbury, which was at one time very prevalent in both breeds, and is even now not altogether banished from either. But it is now fully recognised that canary plumage, deep orange bill and legs, and erect canoe-like body make the true type of the Pekin. It differs further from both the preceding in having no keel.

Indian Runner Ducks

Of the various breeds of ducks, the greatest forager and most prolific layer is beyond doubt the Indian Runner. We are indebted to Mr J.W. Walton, of Tow Law, Co. Durham, for the following notes on the breed:-

Figure 4.1 Aylesbury and Cayuga Ducks

"The best and most authentic account of the origin and introduction of this remarkable breed is that given in the 'History and Description of the Indian Runner Duck', by Mr J. Donald, of Wigton, published about 1890, wherein he states that the first were brought from India about fifty years earlier by a sea captain, who, when ashore, had been attracted by their peculiar carriage and active habits, and after learning of their great egg-producing powers, and that they practically foraged for their living without being artificially fed, brought a few home as a present to some farmer friends in Cumberland. They rapidly established a local reputation for egg production in the district, and seem to have gradually spread northwards over the Scottish border into Dumfriesshire and southwards into Westmorland, but they remained practically unknown anywhere else until about the year 1890.

"At the time Mr Donald wrote they had lost a good deal of their striking, original character, for he remarks that 'very few of the original type are now to be found, and the carriage is not so penguin-like as formerly'; whether owing to climatic influences, to the introduction of foreign blood into many strains, or to in-breeding, it is difficult to say with accuracy, but probably they have all shared to a greater, or lesser degree in producing these defects.

"He also noted the fact that a good many of the Indian Runners of that period had been produced by the use of Runner drakes with farmyard ducks, for we are told, as the fame of the breed as layers extended, the drakes were eagerly sought after and largely employed to cross with the common ducks of the county with a view to improving their egg-producing qualities, the Runner drakes stamping their distinctive characters most pronouncedly upon the produce of the cross, and in a good many so-called Runners seen even then the distinctive characters of the breed were in a great measure obliterated; and it is undoubtedly to the timely publication of Mr Donald's brochure that the Fancy owes the preservation of many of the best and most distinctive characters of the original stock."

Cayuga Ducks

The black Cayuga duck is called after the lake of that name, and comes to us from America, though a large black duck which bred pretty true was known half a century ago in Lancashire. The first American specimens were sent to us by Mr W. Simpson in 1871, and the late Mr J.K. Fowler imported them a few years later. These early specimens were not very large, and were rather dingy in colour, and there is no doubt that they were crossed with Black East India ducks in order to get the green gloss of the latter. This was accomplished, but kept them still small; and they were afterwards crossed, by some with Aylesbury and by others with Rouen, to get size. Unfortunately with this the type was also changed, as the original birds had no "keels" while the modern English exhibition Cayuga has this feature very pronounced. It has thus been made an exhibition duck at the expense of popularity in the market.

Owing partly to this change, perhaps, the Cayuga has never quite had its deserts; for general consent attributes to it decided superiority in flavour over any other of the *large* breeds. It is now a large breed, very similar in shape to the Aylesbury, the plumage being a rich black, heavily glossed with green, the legs a sooty orange, the bill a leaden or bluish black, with an intense black splash in the middle and a black bean at the tip. The skin is very white. The breed is hardy, matures early, is a very good layer, and of more quiet and stay-at-home habits than most. If ever Cayugas should come into fashion, these qualities would be in its favour as a market duck, but the black feathers would, of course, be against it, and may be a reason why it is so little bred. After the first year Cayugas are apt to moult more or less white feathers, especially at the base of the bill; but this is no sign of impurity of race.

The Crested Duck

The following notes on the Crested Duck are from the pen of Mr Scott Miller, Hon. Secretary of the Crested Duck Club:-

"In general appearance this breed is much like the Aylesbury, though not quite so large (adults usually weighing about 8 lbs) and without the heavy keel of the latter; also the head is adorned with a crest or top-knot, which should be globular in shape, as large as possible, and set evenly on the head: a common fault, especially in ducks with large crests, being that the crest hangs more to one side than the other. The breed is to be found in certain districts all over the British Isles, and is also known in America and other places abroad. It is known to have existed in Scotland at least seventy years ago. As to origin, we can only theorise, the theories we have heard advanced being that the breed was introduced pure from the East, or may have been produced by ducks frequenting quiet lakes mating with crested wildfowl, such as Grebes, if such mating be possible, or may have been bred from sports from common ducks; the latter is, in our opinion, the most feasible.

"Whilst being decidedly ornamental, this variety is also a first-class utility one, the ducks being excellent layers of large eggs and themselves making a very creditable appearance on the table. Crested ducks always throw a number of crestless ducklings, but these can be picked out as soon as dry; and, similarly, the size of crest, in crested specimens, can be readily told, as the crests are of the same size, *in proportion,* as they will be when the bird is full grown. This only holds good for the first three or four days after hatching, as the bird soon commences to outgrow its crest, which grows little till the bird gets its first feathers. Fanciers will find that fewer crestless birds hatch year by year.

"It is a strange fact that those youngsters with very large crests generally appear very delicate for the first three days or so, and many die during this period; they rush about apparently afraid of their large head-dress. When this occurs, the breeder should try the remedy of clipping the fluff of the crest short, care being taken not to injure the scalp. In breeding, it is advisable to use crested birds on both sides, though good ones can be bred from plain-headed birds, provided they are crest-bred — that is, from crested parents; but in our opinion this encourages the plain-headed element, which is most objectionable. Choose nice, large, typical birds — one often reads in poultry books that the female has most influence on the size of the progeny, but our experience in this breed is that the drake influences the size quite as much as the duck — with nice, evenly-shaped, and well-placed crests; those having long feathers in the crest are preferable to those with short, the reason being that the crest in ducks, like that in Polish fowls, is composed of a spherical protuberance of the skull-bone, this being the main cause of the crest, length of feather greatly adding thereto. Crested ducks may be of any colour, and need no special care in rearing (beyond what is mentioned above) or general management, being hardy and reaching maturity at an early age. Care should, however, be taken that the crest feathers do not damage the eyes, either by clipping the offending feathers or binding with an elastic band."

Blue Ducks

Blue races of duck are obviously allied to black, and have often appeared. The late Mr Teebay several times told us that about 1860 there was a recognised local race of large blue ducks in Lancashire, and more or less of that colour would be produced by crossing white with either black, or even any dark breed like the Rouen. The same colour has been imported and bred in the United States under the name of blue Swedish ducks, which it is said really were introduced from northern Europe, and only differ from the foregoing in having a white throat or semi-collar at the front of the base of the neck. More recently blue ducks have been bred and sold as "Orpingtons".

Buff and Blue Orpington Ducks

Writing to *The Feathered World* in 1910, Mr A.C. Gilbert, Hon. Secretary of the Orpington Duck Club, thus describes the origin of the Buff and the Blue Orpington duck:-

"Buff Orpington ducks were made some twelve or fourteen years back, but not brought into prominence in England until 1908, when they were shown at several large shows, such as the Dairy and Crystal Palace, where classes were guaranteed for them by Mr A.C. Gilbert, who also at the same time drew up standards for them and formed a club for them. In Australia they had previously made their reputation by winning two twelve months' laying competitions in succession.

"They were made to fill the demand for a first-class layer, and at the same time a nice-sized table bird of fine quality and flavoured flesh. The start was made by mating Indian Runners to Aylesburys, Indian Runners to Rouens, and Indian Runners to Cayugas.

"These different varieties previously to this mating up had been carefully mated and bred for three years for egg-production alone, all laggards in this respect being noted and weeded out, so that when the pure stock was put together for the first crossing a line of good layers was included in all four breeds. The descendants of these matings were then crossed back and forth on to one another until, with time, care, and patience, the desired end was accomplished. But before this came about it was noticed that a great many of the offspring came with a lot of blue on the back, wings, thighs, and portions of the neck. This put the idea into the head of the originator (the late Mr William Cook) to make a Blue Orpington duck, as well as Buff Orpington ducks; so the darkest and the ones with most blue markings were put on one side and mated with Cayugas and Pekins. Then the same process went on with these, mating and intermating, until the Blue Orpington ducks were a finished article, the same as the Buffs.

"Both varieties are most useful and at the same time very handsome and good to look at. They are very hardy, easy to rear, quick in growth, and good foragers on an open space or field, finding most of their own food; very active and fertile, being equally good with a pond or stream, or merely a trough or drinker, for their water. In mild countries, such as the British Isles, they require no houses, doing well out in the open run, field or stackyard."

Campbell Ducks

A successful attempt to create by crossing and selection a new breed of ducks which should exhibit real superiority in useful qualities has resulted in what are known as Campbell ducks, produced by Mrs Campbell, of Uley, in Gloucestershire. These are now bred in two colours. The original strain was descended from one duck which exhibited most remarkable laying powers, and was probably something of the Rouen colour, since the original Campbells are somewhat like Rouens in appearance, but much lighter, with a plain head of a greyish brown shade, and no streak running from the eye: the drakes have grey backs and a pale claret breast — the legs yellow. The object was to produce excellence in laying, with fair table qualities and quick maturity; and it is stated that for years past the egg-average has been over 200 per annum, while the young are hatched at all seasons, and do well all the year round. They are not very large, stock birds weighing 4½ lbs to 5 lbs, and in flavour considerably resemble the wild Mallard, which was used in crossing as one of the foundations of the strain.

The other sub-variety is more recent, and is known as the Khaki or Khaki-Campbell duck. The drake is khaki colour all over except the head and stern, which are bronze green; the duck is entirely khaki colour, a delicate lacing of darker buff showing on each feather. The Indian Runner has been used in crossing to produce this variety, and as the result the Khaki duck is of extremely active habits, doing best on a good range, and showing very little desire for swimming — in fact, Mrs Campbell, we believe, only allows them drinking water. At twelve weeks old the ducklings come up to about 4 lbs to 4½ lbs, the laying being about the same average as the other strain. Whatever time of year they are hatched, they are said to commence laying at or before six months old, so that by hatching about three lots, very early, medium, and late, eggs are easily obtained every day in the year.

Muscovy Ducks

The most distinctive race of all is the Peruvian or Muscovy or Musk duck, which alone appears not to be descended from the Mallard. It comes not from the north at all, but from South America; is very different in many characteristics; and the progeny when crossed with other varieties appear to be real hybrids, being decidedly sterile *inter se,* though fertile more or less with either parent strain. The name is derived from an odour of musk which pervades the skin, but which disappears when cooked. The generic name is *Cairina moschata.*

The wild Muscovy duck is very agile, often perching upon trees, and even making its nest occasionally in such situations. Another peculiarity is the disparity in size between the sexes, a fine drake weighing perhaps 11 lbs or 12 lbs, while the female will be only 6 lbs or 7 lbs; and the male has no curled feathers in his tail like other breeds. The feathers on the body are very large and broad, and often appear loose as if ready to drop out. The head of the drake is very large, and in both sexes the cheeks are naked, with scarlet fleshy carunculations, very developed in the male, and giving him a peculiar leery and wicked expression. This is not belied by his temper which is very bad with other ducks and poultry, and the drakes also fight fiercely among themselves, another point in which they differ from other breeds. The period of incubation also differs, being from thirty-four to thirty-five days.

The general colour of the Muscovy duck is pied black and white; whole white, whole black, and blue dun being also found. The legs are pale yellow, and the toes have very sharp claws. The eggs are large and white, but the duck is a poor layer. The flesh is rich, and at one time the bird was popular in America, and considerably used for crossing; but this has been nearly abandoned, and the breed cannot be considered a profitable one.

GEESE

The Embden Goose

The most valuable breed of geese at the present day, for reasons stated presently, is probably the Embden. The following article on Embden geese was kindly contributed by the Hon. Sybil and Florence Amherst, who were until recently well known as successful breeders and exhibitors of this variety.

"The development of the white domestic goose can be traced from the earliest times. Before show Standards required certain geese to be 'spotless white throughout', it had been the study of breeders, from the remotest ages, for utility purposes, to establish white varieties of geese. An account is given on a papyrus of an Egyptian prince, who, in ordering ten geese to be given as payment to his workmen, cautions those who are to kill the geese not to touch 'the white bird on the cool tank'. In much later times, large flocks of white geese were driven to Rome from north-western Europe for the sake of their feathers. According to Lucretius, the sacred geese that saved the Capitol were white. Horace describes a famous Roman dish made of the liver of white geese fed on fat figs. Varro, about 50 B.C., urges that geese chosen for stock should be large and white, for the goslings are generally like them, and points out the advantage of their 'domesticated, placid nature'. Columella, in the beginning of the Christian era, writing on the same subject, remarks that 'care must be taken that male and female of the largest bodies and of a white colour be chosen' (avoiding the wilder grey kinds). These scattered notes show that it has been by careful selection that pure white breeds of geese have been formed.

"The work of generations is shown to perfection in the Embden goose as now bred and exhibited in the United Kingdom. All nations recognise the creation of this beautiful bird by the English. In Germany, the original home of the Embden, they say, 'Embdens were exported to England a long time ago, and how admirably they have succeeded, and surpassed us with this variety, is well known'. The present race is called 'the new English breed', and as it exists now, is not known in Germany, except as imported from England; and, 'For the last ten years there has been no chance of obtaining a prize at any exhibition for the old type of Embden'. English exhibitors, who have seen the older German Embdens at shows on the Continent, also say, 'They are correct in shape, but too small to compete with our birds'.

"The old Continental Embden goose, or as it is spelt in Germany, Emden (*Anser dom. frieslandicus*), is sometimes, though wrongly, called the 'Bremen' goose. It derives its name from the town of Embden, having been bred in East Friesland, in the valley of the River Em, and the adjoining Jeverland, from time immemorial. The export of feathers from East Friesland to the Levant in very early days formed an important trade, and the rearing of geese on the coast, in the regions of the River Em, for centuries was extensive; but the area of its activity from floods and other causes gradually decreased, and it is now only carried on in two little villages, Riepe and Simonswolde, two or three hours distant on foot from Embden. At the present day, in the middle of May, Embden goslings of four to five weeks old are sent from that district to all parts of Germany, and also to other countries, principally Hungary, Bohemia, and Russia. Eggs for sitting are also sold at high prices, and there is a large export of feathers.

The Toulouse Goose

The Toulouse breed takes its name from the well-known city in southern France, round which birds generally similar to, though not so fine as the English stock, are still reared to a large extent; and it is this variety which is used in the production of the celebrated *pâté de foie gras,* so much imported from the Continent, the essential part of which consists of goose-livers potted with truffles. Did people realise how this delicacy is produced, it is to be hoped that it would be less popular than it is, even among fashionable epicures. To make a painful story as short as possible, even by the more merciful feeders the geese are confined in a very hot room or caged near a stove, and there forced with fattening food until they would die in a day or two more, when they are killed, and their livers are found swelled to an enormous size; but there is unfortunately no room for doubt that by some the wretched birds are tied down to prevent their moving, and by a few, actually nailed by their feet to a board.

This goose is beyond doubt the result of breeding and feeding up the grey or dark descendants of the Grey-lag, and presents marked differences from the preceding in several respects. For the following

Figure 4.2 Toulouse and Embden Geese

descriptive notes we are indebted to Miss Campain, of Deeping St. Nicholas, Spalding, known as a prize-taker with this variety for many years:-

"It is many years since I started breeding geese, and for several years I have been an exhibitor at the leading shows of both the Embden and Toulouse, with a fair share of success, but I treat here of the Toulouse variety. I started by buying a pair of his celebrated geese from Mr Fowler, of Aylesbury, who had then quite as good birds as anyone, if not the best at that time. The gander, I think, was without exception the longest bird in every way I have seen, and the goose was remarkably good in colour, very wide and deep, and not showing the least tinge of brown in plumage, but of a beautiful silver grey.

"This variety should be massive and heavy in appearance in every way. In both the gander and the goose the head should be broad and deep in face, the beak being in a straight line from the top of the head to the tip, very strong and without any indenture or hollow in the top bill, which gives the bill a most objectionable snipy appearance. The bill should be of a rather brown-flesh colour, the dewlap should hang well down, and be as large as possible. The neck should be long and graceful; a short neck completely spoils the appearance, and in mating up for breeding, care should be taken in this point, because they are apt to breed short. Both for the show-pen and for breeding, the birds should be exceptionally well bowed in front and 'keeled' deeply, points which in this variety are of almost primary importance, with their bodies almost touching the ground behind. They must be very broad across the back, and long to the tip of the tail. Their legs must show as large an amount of bone as possible, to get which is a great indication of size and massiveness. In colour they should be rather dark grey on the head, neck, back, and wings; rather lighter on the breast, gradually becoming lighter towards the belly, where it ends in good pure white. The plumage should be as free as possible from a brown tinge, which I strongly object to, though it is prominent in some strains. The sun has some influence on this, but very litte comparatively on good-coloured ones compared to others. The legs are a deep orange colour.

The Chinese or African Goose

The Chinese or African Goose *(Anser cygnoides)* has also been termed the Hong Kong goose and Knobbed goose, and some birds have been written of as Spanish geese, which appear to have had the same general characteristics. It was classed by Cuvier actually with the swans, which it resembles in the longer and more slender neck, and the knobbed bill, also in the neck feathers being smooth and not curled as in the two preceding varieties; and it has been recorded on two or three occasions (on rather doubtful authority, but the gander is such an ardent breeder that it is not unlikely) to have produced swan hybrids. But that it is a true goose is proved by not only its domestic habits and prolificacy, and the number of its vertebrae (sixteen), but by the fact that it breeds freely with other geese, and that the produce is fertile and not a hybrid; the common goose of India being, as Mr Blyth pointed out long ago, a cross between the Chinese and the ordinary domestic race known to us. This immemorial crossing, in India and elsewhere, is the explanation of differences that seem to have puzzled some writers in America, between the "African" goose as there known, and their smaller Chinese.

The orginal Chinese variety ranges over all China, much of Siberia, and most of India, but chiefly northern India. In size it is midway between the wild goose and the swan, but considerably less than our large domestic geese. The neck is long and slender, but the head rather large for the bird, with a knob or protuberance much like that of a swan at the base of the upper bill, and a heavy dewlap under the throat. The usual colour is brownish grey on back and upper parts, passing into light grey or almost white underneath, the breast and front of the neck a yellowish grey, and a dark brown stripe running all down the back of the neck: in this colour the knob is generally black, the bill orange or dark brown, or even black, the legs orange. There are also white birds, which have orange knob and bill; in these also there is a stripe behind the neck which, although white, is more glossy, and different in appearance from the other plumage.

This Chinese race is very prolific, laying twenty to thirty eggs in a sitting, and several sittings in a year, the eggs being about two-thirds the size of those laid by fine ordinary geese. The breed is very hardy and easily reared, and the flesh of very delicate quality; but it is not so domestic in habits as the European goose, and rather fond of swimming at night. It has a harsh and peculiar cry, the most shrill amongst any of the true geese known in domestication.

The "African" goose, as known in America, is stated by American writers to have come from either India or Africa, and is considered by most of them to be a distinct "pure" variety. It is very much larger than that just described, adults weighing as much as 24 lbs for ganders, and 19 lbs for geese, and being described by Mr Cushman and others as actually the largest of all the geese. It is standardised at the same weights as the Embden and Toulouse; but the standard American weights for these are only 20 lbs and 18 lbs for the two sexes, which is far less than in England. All the known facts and circumstances point to the conclusion that this African goose of America is originally simply a cross of the Chinese with the domestic goose, and especially with the Toulouse. During the last thirty years Africa has been opened up in all directions — north, south, east, and west — as it never was before, but no wild goose resembling this breed has ever been found. It probably did come from India, where such crosses have existed for generations; and it is even quite likely that some of these Indian birds may have been carried to East and South Africa by the coolies and Banian traders who have visited those districts so largely; but every single point about the bird tells unmistakably of such a mixed origin. The breed has more of the solid carriage of the Toulouse; the brown has become more grey; the knob is less prominent in proportion; the neck is shorter and thicker; and — most significant of all — the eggs have become fewer, and often as large as those of the older domesticated varieties. The voice also is deeper and more approaching that of the domestic goose, and the flesh more of the same character as in that breed. In the only American specimens of "African" geese which we have seen, a strong element of the ordinary domestic goose was quite unmistakable. It is not necessary to suppose any recent cross-breeding, as some of the Indian birds may quite possibly have possessed all the present features of the African; but it is significant that energetic crossing with the European breeds is now openly practised and strongly recommended by American writers, and it is probable that it has taken place on many occasions. The illustration herewith was given in an American poultry journal as one of the "Chinese" goose, and originally procured for this work as such; but on finally tracing it back, with considerable trouble, to its origin, we found that it really represents a photograph of the "African", in which character it is reproduced here.

This fact illustrates the direct connection between the Chinese and African goose; but it cannot be questioned that in the American modified race, however produced, we have the most valuable and useful form of the breed. Even the African, however, varies considerably. While Mr Cushman places it first in size, and states that it lays the largest eggs of any, but does not put the number higher than in the Embden, if as high, Mr Rankin does not put size so strongly, but states that his birds are better layers than ever, and reach about sixty eggs per annum. These differences are largely accounted for by the fact that recent American breeders have chiefly crossed the African goose, when crossed at all, with the Embden, in order to get as many white goslings as possible.

Figure 4.3 African Geese

The great merit of this characteristic race, however, is as a *breeder*. Every goose breeder knows that the ordinary goose is slow to mate, and requires time, as presently mentioned. Both Chinese and African ganders mate earlier and more quickly, and at a pinch will mate successfully with one or two more geese, being ardent in disposition. The curious fact is also noticed that even an Embden gander will be prolific earlier, and mate with more African geese than with his own variety. As the cross makes weight early, and is of good quality, these facts are of value. There is of course nothing remarkable, to English notions, in ganders of 20 lbs to 24 lbs; but as these are fully equal to the American weights for other breeds, the real aspect of the matter is the existence of a breed equal to Embdens and Toulouse in size, with the advantages just stated; and from this point of view the African, as developed in America, appears well worthy of attention from English goose-breeders who desire an early market. As before intimated, the cross-bred produce are not hybrids, but simply crosses, and perfectly fertile.

The Canada Goose

The only other breed requiring particular mention is the Canada goose, the ordinary wild goose of Canada and the United States, but which has a much wider distribution. It has often been shot on the wing in England, where quite wild flocks of it have been seen; and it ranges through most of the Arctic regions, at least as far north as Spitzbergen. This goose used to form an important portion of the food of the Hudson's Bay Company's trappers, one goose being reckoned as a day's ration, and recorded in the Company's annals as averaging about 9 lbs. The size and weight are thus equal to those of the wild Grey-lag, though in comparison with domestic geese it must be called a rather small breed. The somewhat long body, the long and slender neck, and the character and shape of the head, much resemble those of the swans. Buffon states, indeed, that at Versailles in his time the domesticated Canada geese had bred, or hybridised with the swans kept there, and the older naturalists gave this goose the generic name of *Cygnopsis Canadensis;* but its affinities are clearly with the geese rather than with swans, and it will breed with the races previously described, though the progeny is perfectly sterile.

The head, bill, and greater part of the neck of this goose are black, with a conspicuous white cravat rather than collar at the throat, the head and bill being long. The feathers of the upper parts of the body are greyish-brown, rather lighter at the edges, shading into ashy grey at the wing-coverts, and gradually shading into greyish-white on breast and under-parts, to pure white on the abdomen. The flanks are pale grey tipped with white, the quills of the wings, and tail, almost black. The legs are rather long, set somewhat back so as to give a commanding carriage, and in colour blackish-grey to black. The amount of brown differs somewhat, some birds being an almost pure black and grey.

Figure 5.1 The Common Guinea Fowl

CHAPTER 5

GUINEA FOWL AND TURKEYS

GUINEA FOWL

Under the general head of Guinea fowls, or the genus *Numida,* naturalists have grouped many so-called varieties; but it is very doubtful whether these are even so distinct as the various races of ordinary domestic poultry. The fact that all belong to some part of Africa alone makes a common origin almost certain; and although there is no doubt that the various kinds breed true, that is no more than can be said of the Spanish, Poland, or other common domestic fowls. Many of these sub-races have been crossed, and we believe in every case the progeny have proved fertile, which most naturalists consider evidence of at least close identity; though the whole question of species and what constitutes species needs much more in the way of investigation than it has ever yet received.

Of these various races of Guinea fowls, some have a peculiar bony helmet on the top of the head, while others have this replaced by a crest of feathers, the shape and size of which crest varies in different so-called varieties; and in a third variety which does appear to have some real distinctiveness, there is neither crest nor helmet, and such a resemblance to the vulture generally that the bird has been graphically termed the Vulturine Guinea Fowl. Of the first or helmeted group Mr Elliot and other naturalists have described some half-dozen varieties, but in our opinion several of these are practically identical. The Common Guinea Fowl of West Africa, or *Numida meleagris,* has for a long time been regarded as the original of our domestic race, though some authorities lately have objected to this view, on the ground that as the bird is admitted to have been known to the Romans, and they had more intercourse with the Egyptian side of the great African continent than with the western, one of the varieties common in Abyssinia is more likely to have been the original. We think the common view by far the more likely to be correct. Not only is the name entitled to some weight in a case of this kind, but when residing at Bristol, which is a considerable centre of the West African trade, we have on several occasions seen Guinea fowls perched on the rigging of African vessels, which had been brought from the coast by sailors; and in every case these birds were obviously identical with the domestic breed, both in head and plumage.

The Vulturine Royal Guinea Fowl, as it is called, certainly does present very peculiar and singular characteristics. The neck and tail are very long in comparison with the common variety, and the other points have been described as follows:- The head and upper part of the throat are destitute of feathers, but sprinkled with hairs of a black colour, which are longest on the neck; the nape is thickly clothed with short, velvet-like, brown down; and the lower part of the neck ornamented with long, lanceolate, and flowing feathers, having a broad stripe of white down the centre, to which on each side succeeds a line of dull black, finely dotted with white, and margined with fine blue. The feathers of the inferior part of the back are of similar form, but broader, with a narrower line of white down the centre, and with the minute white dots disposed in irregular and obliquely transverse lines. The wing-coverts, back, rump, tail, under tail-coverts, and thighs, are blackish brown, ornamented with numerous round and irregular spots of white surrounded by circles of black, the intermediate spaces being filled by very minute spots of dull white; the primaries are brown, with light shafts and spots of brownish white on the outer web; the secondaries brownish black on the tips, with three imperfect lines of white disposed lengthwise on the outer web, and three rows of irregular spots of white on the inner web; the breast and sides of the abdomen are of a beautiful metallic blue, the centre of the abdomen black, the flanks dull pink, with numerous spots of white surrounded by circles of black; the bill is brownish, and the feet brown.*

Mr Gould writes of this magnificent variety of Guinea fowl:- "Independently of the chaste and delicate markings which adorn the whole of this tribe, the neck of the present species of Guinea fowl is ornamented by a ruff of lanceolate flowing plumes; which new feature, as well as the head being entirely devoid of fleshy appendages, renders it conspicuously different from all its congeners. It is certainly one of the most noble birds that has been discovered for some years; and we indulge in the hope that the period may not be far distant when we shall become better acquainted with the species, and that living individuals may even become denizens of our menageries and farm-yards, where they would doubtless thrive equally well as their congener so familiar to us all".

It is the long neck adorned with lanceolate feathers, the absence of casque or crest, and the long tail and legs, which give this bird so strange a resemblance to the vulture. It has been introduced into one or two menageries; but there is little probability as yet of its being introduced as farm stock. We may here repeat our remark in the last chapter, that even the most intelligent keeper of a zoological institution is not the most likely person to multiply and naturalise a new race of poultry; his knowledge is not special enough, and in this particular walk is far surpassed by that of any intelligent and enthusiastic poultry amateur. Instead of merely giving prizes for rearing, as was once done in relation to curassows, a zoological society would act in a manner far more likely to attain its object, were eggs or stock of the desired new variety to be given to such amateurs of skill and standing as were willing to accept them, and had at command the ample space they so imperatively require; when the most special care would be given, and all that skill, intelligence, and enthusiasm could suggest would be cheerfully lavished upon the new breed, in a way which no public institution can ever secure.

The domestic Guinea fowl in ordinary circumstances can hardly be considered profitable poultry, but its character has nevertheless been considerably belied. We have heard it said that it could not be kept on account of the screeching noise it makes; but we cannot understand how any one objecting on that ground can abide the noise of an ordinary cock, much less of an ordinary farm-yard; since, disagreeable as the cry is (resembling the noise of a creaking axle more than anything else we can think of), it is very seldom heard near the house. Sounder objections are found in the straying proclivities of the fowl; its disposition to lay away, by which many eggs are lost; and its pugnacious habit of beating other varieties of poultry. But for this latter trick it would long since have been naturalised as a game bird, having been turned into covert with perfect success; but it was found that the Guinea fowls drove away other descriptions of game to such an extent that the birds had to be destroyed on that account, the pheasants being most valued.

TURKEYS

Passing from the wild Turkey to the domestic bird and its management, we may observe that this too seems to have merged into three tolerably marked and definite varieties, known as the Norfolk Turkey, the Cambridge, or variegated variety, and the beautiful "bronzed" Turkey recently introduced from North America.

The counties of Norfolk and Cambridge have long been celebrated for the immense number of turkeys they send to the London market, and which constitute a trade as well marked as the poultry-raising which we have already described as carried on in Surrey and Sussex. As a few particulars in this case also may prove both useful and interesting, we extract the following remarks from a paper by Mr H.H. Dixon in the Journal of the Royal Agricultural Society:-

"The eastern counties," he says, "may be said to have pretty nearly a monopoly of our English turkey raising and feeding. Hen-wives are generally 'afraid to meddle with them', on the score of delicacy; but if the requisite food and attendance are not found to be thrown away in Norfolk, Cambridgeshire, &c., why should they be elsewhere, except in an essentially damp climate? They must be tenderly reared, and not 'dragged up', as the saying is.

"The Norfolk Turkey is black, with a few white spots on its wings. The Cambridgeshire Turkey is of a bronze grey, and rather longer in the leg and bigger in the bone. Very few white ones are to be seen, as they are supposed, like a white long-horn cow, to be more delicate.

* *"Cassell's Book of Birds," Vol. III., p.256*

Figure 5.2 Wild Turkeys at Home

CONCLUSION

What an exciting era the Victorians lived in! The energetic stimulus must have been quite remarkable. From birds imported from various countries a complete Fancy was developed to a level not achieved in modern times.

The multitude of breeds, and varieties within these breeds, indicate the size of the task undertaken by the pioneers of the poultry fancy. Moreover, when a breed became popular the effort put into achieving perfection was quite incredible. Look at the large Modern Game Fowl which was developed to a point when probably no further improvement was feasible. Examine the double-mating used with certain breeds to develop distinct characteristics in males or females. Imagine the problems that were encountered in stabilising the colours, combs and other desired characteristics. What patience and dedication were necessary to achieve the desired ends.

We can but marvel at the results. Hopefully we can also learn from what has gone before. Tracing the history of a specific breed can be a rewarding task. It is hoped that the study of **Lewis Wright's Poultry** *will inspire others to carry on the delightful hobby of poultry keeping.*